BUDDHISM & SCIENCE

BUDDHISM AND MODERNITY

A series edited by Donald S. Lopez Jr.

RECENT BOOKS IN THE SERIES

Critical Terms for the Study of Buddhism, edited by Donald S. Lopez Jr. (2005)

The Madman's Middle Way: Reflections on Reality of the Tibetan Monk Gendun Chopel, by Donald S. Lopez Jr. (2006)

The Holy Land Reborn: Pilgrimage and the Tibetan Reinvention of Buddhist India, by Toni Huber (2008)

BUDDHISM & SCIENCE

A Guide for the Perplexed

DONALD S. LOPEZ JR.

THE UNIVERSITY OF CHICAGO PRESS

Chicago and London

DONALD S. LOPEZ JR. is the Arthur E. Link Distinguished University Professor of Buddhist and Tibetan Studies in the Department of Asian Languages and Cultures at the University of Michigan. He is the author or editor of a number of books, including *The Madman's Middle Way: Reflections on Reality of the Tibetan Monk Gendun Chopel; Prisoners of Shangri-La: Tibetan Buddhism and the West; Critical Terms for the Study of Buddhism*; and *Curators of the Buddha: The Study of Buddhism under Colonialism.*

The University of Chicago Press, Chicago 60637
The University of Chicago Press, Ltd., London

Printed in the United States of America

17 16 15 14 13 12 11 10 09 08 1 2 3 4 5

ISBN-13: 978-0-226-49312-1 (cloth)
ISBN-10: 0-226-49312-1 (cloth)

Library of Congress Cataloging-in-Publication Data

Lopez, Donald S., 1952–
Buddhism and science : a guide for the perplexed / Donald S. Lopez.
p. cm. — (Buddhism and modernity)
Includes bibliographical references and index.
ISBN-13: 978-0-226-49312-1 (hardcover : alk. paper)
ISBN-10: 0-226-49312-1 (hardcover : alk. paper)
1. Buddhism and science.
I. Title.
BQ4570.S3L67 2008
294.3'36509—dc22

2008012963

♾ The paper used in this publication meets the minimum requirements of the American National Standard for Information Sciences—Permanence of Paper for Printed Library Materials, ANSI Z39.48-1992.

In Memory of My Father

DONALD S. LOPEZ

1923–2008

How easily these old worships of Moses, of Zoroaster, of Menu, of Socrates, domesticate themselves in the mind. I cannot find any antiquity in them. They are mine as much as theirs. **RALPH WALDO EMERSON, 1841**

CONTENTS

PREFACE

In the winter of 1870–71, Ernst Johann Eitel (1838–1908), a member of the London Missionary Society, delivered a series of lectures on Buddhism at the Union Church in Hong Kong. Eitel was one of the great missionary-scholars of the Victorian period, an accomplished sinologist who also read Sanskrit. His ultimate goal was to demonstrate the falsity of Buddhism. Yet in his third lecture, he enumerated some of the ways in which Buddhism had anticipated science:

> Though no Buddhist ever attained to the clearer insight and mathematical analysis of a Copernicus, Newton, Laplace or Herschel, it must be acknowledged that Buddhism fore-stalled in several instances the most splendid discoveries of modern astronomy. Teaching the origin of each world to have taken place out of a cloud, the Buddhists anticipated 2,000 years ago Herschel's nebular hypothesis. And when those very patches of cloudy light or diffused nebulosities which Herschel believed to be "diffused matter hastening to a world birth" dissolved themselves before the monster telescope of Lord Rosse into as many assemblages of suns, into thousands of other world-systems dispersed through the wilds of boundless space, modern astronomy was but verifying the more ancient

Buddhistic dogma of a plurality of worlds, of the co-existence of thousands of chiliocosmoi inhabited by multitudes of living beings.[1]

Eitel invokes five great names in the history of astronomy: Nicolas Copernicus (1473–1543), whose *On the Revolutions of the Celestial Spheres* (*De revolutionibus orbium coelestium*) presented the heliocentric theory of the universe; Sir Isaac Newton (1643–1727), who invented the refracting telescope and explained the role of gravity in planetary motion; Pierre-Simon Laplace (1749–1827), who developed mathematical methods for calculating and predicting the motion of the planets; William Herschel (1738–1822), discoverer of Uranus and cataloger of nebulae; and William Parsons, third Earl of Rosse (1800–1867), who in 1844 built the "Leviathan of Parsonstown," the world's largest telescope. Each of these figures would have been well known to Eitel's expatriate audience in the Hong Kong church.

Laplace and Herschel were associated with the nebular hypothesis, a theory previously propounded by both Emanuel Swedenborg and Immanuel Kant, which postulated that a solar system originated from a mass of incandescent gas—for Herschel it was a shining fluid that he called "true nebulosity"—rotating on an axis, eventually contracting into a mass. The outer rings of this mass broke off to form planets, with the central core becoming their sun. One of the great debates in astronomy in the nineteenth century was whether this incandescent fluid indeed existed or whether it was instead a mass of distant stars. In early 1846, Rosse and his monster telescope showed that the Orion Nebula could in fact be resolved into stars.

These were some of the latest scientific discoveries of Eitel's day. And he claims that they have been "forestalled" (by which he means "anticipated") two thousand years ago by the Buddhists. Eitel is referring to a Buddhist account of the origin of the world. Faint winds, impelled by the force of karma, begin to blow in the vacuity of space, eventually converging to form a circle of wind, described as solid and indestructible. A thick cloud forms above the circle of wind, raining down drops of water of various sizes that together become a great ocean, supported on the circle of wind. In this ocean, a thousand golden lotus flowers appear. The churning of the ocean eventually gives rise to a ring of mountains that contains the waters. In the center of the ocean, a great mountain appears, with an island (flanked by two smaller islands) in each of the four cardinal directions. This is a world, and a thousand of these worlds is a Buddhist universe, what Eitel calls a "thousand world" or chiliocosm.[2]

Eitel sees in the Buddhist rain cloud an anticipation of Herschel's nebulae, and in the Buddhist "thousand world" an anticipation of galaxies, anticipated without the assistance of Rosse's giant lens. These worlds were inhabited by "multitudes of living beings." Eitel, in keeping with the views of many astronomers of his day, believed that the planets were populated. Indeed, late in life, Herschel had published a paper arguing that the sun was inhabited, with two layers of dense clouds protecting the inhabitants from the intense light of the luminous shell observed from earth; sunspots may be the peaks of tall mountains rising through the shell.[3]

We see, then, a Christian missionary, almost a century and half ago, making grudging claims for the compatibility of Buddhism and Science. Over the ensuing decades, such claims have continued to be made with a remarkable persistence. This book is a study of that persistence.

Its central claim is a modest one. It is that in order to understand the conjunction of the terms *Buddhism* and *Science*, it is necessary to understand something of the history of the conjunction. It might be dated back to the sixteenth century, when Saint Francis Xavier, the Jesuit missionary to Japan, noted that the Buddhists do not understand that the world is round. It might be traced back to the Reverend Dr. Eitel's lectures from his Hong Kong pulpit. Or it might be traced to the year 1873, when the Wesleyan minister David da Silva in Sri Lanka held up a globe during a debate with a Buddhist monk and asked him to locate Mount Meru, the cosmic peak that rose from the waters to form the center of the Buddhist world. That these events occurred in the course of Christian missions to Buddhist Asia suggests that Buddhist claims about Science originated in polemic, with Buddhists arguing that their religion is not superstition but science. Yet such claims have persisted after the opponent in that polemic has disappeared, or has at least become less visible. And the claims of compatibility have not always originated among Asian Buddhists. The discourse of Buddhism and Science has been transmitted through networks that crisscross the nebulous boundaries of East and West. Asian Buddhists have argued for the compatibility in order to validate their Buddhism. European and American enthusiasts and devotees have argued for the compatibility in order to exoticize Science, to find it validated in the insights of an ancient Asian sage.

A second assertion of this book is that for more than 150 years, the claims for the compatibility of Buddhism and Science have remained remarkably similar, both in their content and in their rhetorical form. This similarity has persisted despite major shifts in what is meant by *Buddhism* and what is meant by *Science*. In the early decades of this history, Buddhism generally referred to what European scholars dubbed "original Buddhism," the Buddhism of the Pāli canon, preserved in the Theravāda traditions of Southeast Asia and Sri Lanka. In the period after the Second World War, although the Theravāda continued to be regarded as "Buddhism" in some quarters, Zen came to the fore. And since the 1990s, Tibetan Buddhism has displaced Zen to become the chief referent of Buddhism in the Buddhism and Science dialogue, largely through the influence of the Fourteenth Dalai Lama. Still, over the course of almost a century and a half, the Buddha is said to have somehow anticipated the most up-to-date view of modern science as thousands of pages of the calendar have been turned.

The referent of *Science* is also nebulous. At times, *science* has meant a method of sober and rational investigation, with the claim that the Buddha made use of such a method to arrive at the knowledge of deep truths about inner and outer worlds. At other times, *science* refers to a specific theory: the mechanistic universe, the theory of evolution, the theory of relativity, the big bang, whose antecedents are to be found in Buddhist doctrine. At other times, *science* has referred to a specific technology—the microscope, the telescope, the spectrometer—that has been used to discover what the Buddha knew without the aid of such instruments; as more precise instruments have been developed over the past century, the claims of the Buddha's knowledge have remained constant. And at still other times, *science* has referred to the manipulation of matter, with dire consequences for humanity unless paired with the compassionate vision of the Buddha.

From the traditional perspective, the Buddhist truth is timeless; the Buddha understood the nature of reality fully at the moment of his enlightenment, and nothing beyond that reality has been discovered since. From this perspective, then, the purpose of all Buddhist doctrine and practice that have developed over the two and a half millennia is to make manifest the content of the Buddha's enlightenment. From the historical perspective, the content of the Buddha's enlightenment is irretrievable, and what is called Buddhism has developed in myriad forms across centuries and continents, with these forms linked by their retrospective gaze to the solitary sage seated beneath a tree. From either perspective, in order to make this "Buddhism" compatible with "Science," Buddhism must be severely restricted, eliminating much of what has been deemed essential, whatever that might be, to the exalted monks and ordinary laypeople who have gone for refuge to the Buddha over the course of more than two thousand years.

If something is lost, what is gained? This book surveys the long history of the discourse of Buddhism and Science in an effort to understand why we yearn for the teachings of an itinerant mendicant in Iron Age India, even one of such profound insight, to somehow anticipate the formulae of Einstein.

INTRODUCTION

The religion of the future will be a cosmic religion. It should transcend a personal God and avoid dogmas and theology. Covering both the natural and the spiritual, it should be based on a religious sense arising from the experience of all things, natural and spiritual as a meaningful unity. If there is any religion that would cope with modern scientific needs, it would be Buddhism. **ALBERT EINSTEIN**

If Buddhism and Science had a Bible, it would open with these words, and all the chapters and verses would comment on them. These words are a prophecy, declaring that the religion of the future will not be limited to our small planet but will encompass the cosmos. All religions have sought to do this, but the implication here is that the religion of the future will do so accurately, its cosmology based not on myth but on physics. With scientific insight into the nature of the cosmos, the religion of the future will no longer need the primitive notion of a personal God who created the world and who dispenses rewards and punishments to his creatures. This religion will encompass both the spiritual and the natural in a harmonious way. It will require no confessions of faith or assents to propositions that derive from the authority of a scripture or a church.

Instead, it will be based on experience, on the individual's sense of oneness. And, finally, it will be compatible with science. This will be the religion of the future, but it is a religion that already exists. It is, indeed, an ancient religion. It is Buddhism, set forth by the Buddha, the Enlightened One, over two millennia ago. The power of this prophecy is derived from the authority of the prophet, Albert Einstein, the Buddha of the modern age.

But it seems that Einstein never said this.[1] And on second reading, there is something about the statement that is too good to be true; it is too perfect, too comforting. Or so I will hope to show in the pages that follow.

• • •

Buddhism and Science. What does this term, with its conjunction of two such potent words, imply? The answer to this question depends, of course, on what one means by *Buddhism*, what one means by *Science*, and, not insignificantly, what one means by *and*. However, those who have used the term over the course of more than a century have generally intended some kind of kinship, or at least compatibility, between Buddhism and Science, regardless of what they have understood *Buddhism* and *Science* to be. The nature of this compatibility has ranged across a wide spectrum, with some suggesting that the essential teachings of Buddhism (variously identified) are in no way contradicted by the findings of science (variously enumerated), while others suggest that the Buddha anticipated many of the key discoveries of science, that the Buddha knew more than two millennia in the past what scientists would only discover more than two millennia in the future.

Such claims have been commonly made over the course of the twentieth century and into the twenty-first, as the most cursory glance through the burgeoning bibliography of Buddhism and Science will demonstrate. A random selection of a dozen recent titles (excluding books and articles simply entitled "Buddhism and Science") might include "Time in Madhyamika Buddhism and Quantum Physics," *Psychotherapy and Buddhism: Toward an Integration*, "Quantum Mechanics and Compassion," *Zen and the Brain: Toward an Understanding of Meditation and Consciousness*, "Emptiness and Relativity," *Zen Buddhism and Modern Physics: Morality and Religion in the New Millennium*, "Galaxies and

Śūnyatā," *Two Views of Mind: Abhidharma and Brain Science*, "The Relevance of the Buddhist Theory of Dependent Co-origination to Cognitive Science," "Atom and Anattā," "Karma, Rebirth, and Genetics," *Einstein and Buddha: The Parallel Sayings*, and "A Middle Way: Meditation in the Treatment of Compulsive Eating." A series of books by and of dialogues with the current Dalai Lama has appeared over the past decade.

The claim that an itinerant teacher of Iron Age India understood the theory of relativity, quantum physics, or the big bang theory (each of which has been asserted) would seem to be preposterous. Yet such claims have been made for more than a century, substituting whatever is regarded as the most advanced scientific knowledge of the day as a component of the Buddha's enlightenment. Similar claims have been made by various Hindu revivalists and fundamentalists over roughly the same period, beginning in the mid-nineteenth century and continuing to the present. Thus, it has been asserted that railroads and air travel are reported in the Rig Veda, that the beam of light emitted from Śiva's forehead is a laser, that the opposing armies in the great battle at Kurukṣetra described in the *Mahābhārata* used nuclear weapons. That claims about compatibility with European science began to be made by both Hindu and Buddhist leaders in the same period—the latter half of the nineteenth century—suggests that the encounter with European colonialism played some role in the process. However, the fate of such claims in the decades that have followed has been quite different. The statements about Hinduism are met, at best, with an indulgent smile. Yet the declaration that the Buddha understood the theory of relativity occasions serious reflection. It is the case that the Hindu and Buddhist contentions diverge in their content. The Hindu claims often take the form of an assertion that inventions and technologies believed to have originated in Europe and America in the nineteenth and twentieth centuries had in fact existed in India more than two millennia ago. The Buddhist claims, or claims about Buddhism, tend to be different, focusing not so much on technology but on scientific theory, a knowledge that the Buddha possessed so long ago. Still, this seems insufficient to account for the wide divergence in both the reception and subsequent circulation of Hindu and Buddhist claims. There is clearly something about *Buddhism* that has sustained its long conjunction with the word *science*. This book will try to identify what that something might be.

This book will not attempt to survey the substantial "Buddhism and Science" literature.[2] Nor will it attempt to assess the validity of the wide range of claims made about the compatibility of particular scientific facts and theories on the one hand and various points of Buddhist doctrine on the other. I certainly am not qualified to judge their accuracy from a scientific perspective, even if what might be meant by *accuracy* in this case could be defined and standards for its measurement could be established. The authors of most books and articles about Buddhism and Science generally fall into one of several categories. Some are Asian Buddhist monks (especially from Sri Lanka) with some knowledge of Western science; some are Buddhist monks, or former Buddhist monks, of European or North American parentage, with some previous education (such as an undergraduate degree) or professional training in science; some are Asian scientists from Buddhist cultures who regard themselves as Buddhists; some are European or American scientists with at least a passing interest in Buddhism, an interest that may extend to practicing meditation and identifying themselves as Buddhists. I have none of these qualifications. I write as a historian of Buddhist thought and practice, with an interest in the processes by which what we today call "Buddhism" has emerged in modernity. It is from this perspective that I have approached the subject.[3]

Since the discourse of Buddhism and Science began in earnest in the second half of the nineteenth century, it is perhaps useful to recall how Buddhism was understood at that time. An appropriate place to seek this understanding is the famous ninth edition of the *Encyclopædia Britannica*, which began publication in 1875. The entry on Buddhism was written by Thomas W. Rhys Davids (1843–1922), the most influential British scholar of Buddhism of his day. Here is how he began:

> BUDDHISM is the name of a religion which formerly prevailed in India, and is now professed by the inhabitants of Ceylon, Siam, and Burma (the southern Buddhists), and of Nepāl, Tibet, China, and Japan (the northern Buddhists). It arose out of the philosophical and ethical teachings of Siddhārtha Gautama, the eldest son of Suddhōdana, who was rāja in Kapilavastu and chief of the tribe of Sākyas, an Aryan clan seated during the 5th century B. C. on the banks of the Kohain about 100 miles N. of the city of Benāres, and about fifty miles S. of the foot of the Himālaya Mountains.

We are accustomed to find the legendary and miraculous gathering, like a halo, around the early history of religious leaders, until the sober truth runs the risk of being altogether neglected for the glittering and edifying falsehood. Buddha has not escaped the fate that has befallen the founders of other religions. . . . It is admitted that, under the mass of miraculous tales which have been handed down regarding him, there is a basis of truth already sufficiently clear to render possible an intelligible history, which will become clearer and clearer as older and better authorities are made accessible.[4]

A number of Rhys Davids's points merit comment. He describes Buddhism as a religion; some advocates of Buddhism and Science would later dispute this characterization, calling Buddhism instead a philosophy, a way of life, even a science. He divides the Buddhist world in two. With Buddhism defunct in India, the land of its origin, two Buddhisms remain: the southern Buddhism of Sri Lanka (where Rhys Davids had himself served as a colonial official from 1864 to 1874) and Southeast Asia, and the northern Buddhism of East Asia, Tibet, and Nepal. It was Rhys Davids's strongly held opinion that it was the former Buddhism which was closest to the Buddhism of the Buddha and which more accurately preserved his teachings in its canon. This idea of different Buddhisms would persist, and their number would multiply. In James Hastings's influential *Encyclopædia of Religion and Ethics*, whose first volume appeared in 1908, we find under the heading "Buddhism," "The character of Buddhism varies according to the country in which it prevails, so that a general sketch would be of very little value." And yet, there is a long entry there on the Buddha.

Scholars of this period perceived a great divide between the Buddha and Buddhism. Thus, in the passage above, Rhys Davids turns immediately to the Buddha, describing his teachings as "philosophical and ethical." Buddhism is a religion, but its founder taught philosophy and ethics. Noting that the Buddha was an "Aryan," Rhys Davids provides some sense of what has turned the philosophy of the Buddha into a religion: it is the inevitable admixture of legend and miracle into the life of the man. Following upon the quest for the historical Jesus, Rhys Davids believed that the historical Buddha could be discerned beneath the encrustations of myth, and that he would appear more clearly as more ancient records became available (which has not, in fact, occurred).

The Buddha had lived in northern India in the fifth century BCE. Buddhist monks carried texts and images to much of Asia over the next millennium and a half; the Buddha is famously reported to have instructed his monks to "wander forth for the good of the many, for the happiness of many, out of compassion for the world, for the good, benefit, and happiness of gods and humans. Let no two go in the same direction." By the fourteenth century, Buddhism had all but disappeared from India. The reasons for its disappearance were complex, but its consequences were profound. This demise of Buddhist institutions in India would be a crucial factor in the Buddha's posthumous career in Europe as the founder of a great world religion. For by the time that European scholars (notably those of the British East India Company), trained in South Asian languages, began a sustained study of the culture and history of India, what they would come to call Buddhism was a relic. There were no Buddhists in India, although there were Buddhists almost everywhere else in Asia. Instead there were monuments (often in ruins), statues (often broken), and texts. These were the materials from which European scholars would build Buddhism. Central to this process was the European adoption of the Buddha—a great man, yet still a man—who had been left orphaned by the tradition that he founded, a tradition that had quickly deified him as the sober community of monks that he established yielded to the demands of the superstitious laity on whom they depended for their alms.

The great French Sanskrit scholar Eugène Burnouf, regarded as the founder of the academic study of Buddhism in Europe, wrote in 1844, "I speak here in particular of the Buddhism that appears to me to be the most ancient, the human Buddhism, if I dare call it that, which consists almost entirely in very simple rules of morality, and where it is enough to believe that the Buddha was a man who reached a degree of intelligence and of virtue that each must take as the exemplar for his life."[5] The Buddha was portrayed as a great reformer, described by some as "the Luther of Asia," who condemned the vapid priestcraft of the brahmans and the caste system they controlled. His was a religion, if it was a religion at all, that required no dogma, no faith, no divinely inspired scriptures, no ritual, no worship of images, no God. This view of the Buddha seemed to have enjoyed particular popularity among the more anticlerical of the European scholars. Thomas Huxley, "Darwin's Bulldog" himself, described Buddhism in 1894 in these terms:

> A system which knows no God in the western sense; which denies a soul to man; which counts the belief in immortality a blunder and the hope of it a sin; which refuses any efficacy to prayer and sacrifice; which bids men look to nothing but their own efforts for salvation; which, in its original purity, knew nothing of vows of obedience, abhorred intolerance, and never sought the aid of the secular arm; yet spread over a considerable moiety of the Old World with marvellous rapidity, and is still, with whatever base admixture of foreign superstitions, the dominant creed of a large fraction of mankind.[6]

For Huxley and other Victorians, Buddhism was a tradition that saw the universe as subject to natural laws, without the need for any form of divine intervention. This led many European enthusiasts to declare Buddhism as the religion most suited to serious dialogue with Science, because both postulated the existence of immutable laws that governed the universe. This claim would continue to be made over the next century as one immutable natural law was vetoed and another was ratified in its place.

The nineteenth century was a period of significant advances in the science of philology, with the discovery of language families and ancient connections between the classical Indian language of Sanskrit and the classical European languages of Greek and Latin, as well as modern German, French, and English. These were called the Indo-European or Aryan languages; *āryan* is a Sanskrit term meaning "noble" or "superior," and was the name that ancient peoples of northern India used to refer to themselves. (The fate of the term *āryan* in the Buddhism and Science discourse of the nineteenth century is discussed in chapter 2.) Through a complicated process, theories of language groups gave rise to theories of racial groups; and the kinship between the people of ancient India and the people of ancient Greece and hence (through a certain leap of faith) those of modern Europe was not simply a matter of verb roots but of bloodlines. From this perspective, the Buddha was not so foreign.[7] He was in fact, racially, an Aryan. But the nobility of the prince who had renounced his kingdom was not only hereditary. The Buddha had famously rejected the idea of inherited nobility, claiming that nobility derived instead from wisdom. He thus called his first teaching the four truths for the noble (*āryan*)—not "the four noble truths," as the phrase has been famously mistranslated. The Buddha became, thus, doubly noble. He was noble by birth, by blood, and by language, yet he was also noble because he renounced his royal birth to achieve a spiritual nobility. In a Europe obsessed with questions of race and questions of humanity, the Buddha was both racially superior and a savior for all humanity, an ancient kinsman, a modern hero. The European creation of this Buddha is described in chapter 4.

This Buddha was rather different from the Buddha whose words were recited and whose image was venerated across Asia. As the editors of the *Encyclopædia of Religion and Ethics* had noted in 1908, there were many Buddhisms in Asia, and thus there were also many Buddhas. However, he was also portrayed with a great consistency across the geographical and historical range of the tradition. He was believed to have perfected himself over the course of millions of lifetimes as a bodhisattva, performing the virtuous deeds called perfections (*pāramitā*). He was believed to have taken rebirth in heaven in his penultimate life, where he surveyed the world to select the city of his final birth, his caste, his clan, and his parents. He was believed to have achieved enlightenment at the age of thirty-five, sitting under a tree near the banks of a river. With this enlightenment, and even before it, he was believed to possess all manner of supernormal powers, including full knowledge of each of his own past lives and those of other beings, the ability to know others' thoughts, the ability to create doubles of himself, the ability to rise into the air and simultaneously shoot fire and water from his body. It was said to be an extraordinary body, endowed with the thirty-two marks of a superman (*mahāpuruṣa*), including forty teeth, webbed fingers, the pattern of wheels on the soles of his feet, and a crown protrusion (*uṣṇīṣa*) on his head. He was believed to have passed into nirvāṇa at the age of eighty-one, although he could have lived "for an aeon or until the end of the aeon" if only he had been asked to do so.

Rather than narrating the events of his final lifetime, the traditional biographies of the Buddha, which did not begin to appear until some four centuries after his death, seemed more intent on describing his previous lives, and the previous buddhas he encountered along the way. The concern there seemed to be the portrayal of not the Buddha's individuality and humanity, but his identity with other buddhas said to have come in the distant past, and who were prophesied to come in the future. He was but one of many buddhas; each taught the same truth, the same path to liberation from suffering. There were indeed elaborate descriptions of the Buddha's final birth, but they appeared centuries after his death; and even there, it was his continuity with the buddhas of epochs past, rather than his unique character, that provided the foundation of his authority. Indeed, all buddhas are said to be remarkably similar in word and deed; they differ from each other in just a few ways, one of which is the circumference of their auras.

One need not fault European scholars of the nineteenth century for their attempts to shade their eyes from that aura in order to discern the historical Buddha, for such a person did exist in India in the sixth and fifth centuries (or fifth and fourth; there is uncertainty about the year of his death). Their quest for the historical Buddha differs, however, from the quest for the historical Jesus in important regards. First, unlike in the case of Jesus, no contemporary records or accounts by the first generation of his disciples exist. Like Jesus, the Buddha wrote nothing. But unlike Jesus's followers, the followers of the Buddha wrote nothing. Indeed, the teachings of the Buddha were not committed to writing until the end of the first century BCE, and not in India but in Sri Lanka. Furthermore, there was no Josephus of the Buddha's day, no learned chronicler to provide an account of his life.

Second, the scholars who undertook the quest for the historical Buddha were not products of a Buddhist culture. They were instead from what seemed another world, or so they perceived themselves, writing in the context of colonialism. In 1844 Eugène Burnouf argued convincingly that Buddhism is an Indian religion and that it must be understood first through texts in Indian languages. For the remainder of the nineteenth century, India became the primary focus of Buddhist studies in Europe, and Sanskrit (together with Pāli) became the lingua franca of the field. These were indeed the classical languages of Buddhism. They were also Aryan languages, related to Greek and Latin, the classical languages of Europe. The Buddha of India was thus doubly important, as both founder and forefather. Much of the early scholarship focused on the life of the Buddha and on the early history of Buddhism in India, prior to its demise there, referred to by such terms as *original Buddhism*, *primitive Buddhism*, sometimes *pure Buddhism*. This austere system of ethics and philosophy stood in sharp contrast to what was perceived as the spiritual and sensuous exoticism of colonial India, where Buddhism was long dead. This ancient Buddhism, derived from the textual studies of scholars in the libraries of Europe, could be regarded as the authentic form of this great religion, against which the various Buddhisms of nineteenth-century Asia could be measured, and generally found to be both derivative and adulterated. Buddhism thus came to be regarded as a tradition that resided most authentically in its texts, such that it could be effectively studied from the libraries of Europe; many of the most important scholars of the nineteenth century never traveled to Asia. Relatively little attention was paid to the ways in which the Buddha was understood by the Buddhists of Asia, both past and present.

But this Buddha, created in Europe, did not remain there. As in the classical colonial economy, raw materials—in this case, Buddhist texts in Sanskrit and Pāli—were extracted from the colony and shipped to Europe, where they were refined to produce a new Buddha, one that had not existed before. To complete the colonial circuit, that Buddha was then exported back to Asia, where he was sold to Asian Buddhists at a high price. This is the Buddha who would be hailed by Japanese defenders of Buddhism against the Meiji government, which regarded Buddhism as a foreign superstition. This was the Buddha who was hailed by Sinhalese reformers, like Anagārika Dharmapāla, as superior to Jesus. This was the Buddha of Buddhism in the phrase "Buddhism and Science."

In 1873 a debate took place in Sri Lanka between a Buddhist monk and a Wesleyan clergyman on the respective merits of Buddhism and Christianity. This debate (discussed at length in chapter 1) would have far-reaching effects for the history of modern Buddhism in general, and the discourse of Buddhism and Science in particular. Five years after the debate, in 1878, an account was published in Boston, *Buddhism and Christianity Face to Face*. It was read by Colonel Henry Steel Olcott, a journalist and veteran of the American Civil War. In New York City three years earlier, Olcott and Madame Helena Petrovna Blavatsky, a Russian émigré, had founded the Theosophical Society. The goals of their society were "to diffuse among men a knowledge of the laws inherent in the universe; to promulgate the knowledge of the essential unity of all that is, and to determine that this unity is fundamental in nature; to form an active brotherhood among men; to study ancient and modern religion, science, and philosophy; and to investigate the powers innate in man." The Theosophical Society was one of several responses to Darwin's theory of natural selection during the late nineteenth century. Rather than seeking a refuge from science in religion, Blavatsky and Olcott attempted to found a scientific religion, one that accepted the new discoveries in geology and biology while proclaiming an ancient and esoteric system of spiritual evolution more far-reaching and profound than anything Darwin had described. Indeed, Madame Blavatsky accepted the theory of evolution but rejected claims that life emerged from matter rather than spirit.

By 1878 Blavatsky and Olcott had shifted their emphasis away from "spiritualism" and the investigation of psychic phenomena toward a broader promotion of a universal brotherhood of humanity, claiming affinities between Theosophy and the wisdom of the Orient, specifically Hinduism and Buddhism.[8] And inspired by Olcott's reading of the account of the 1873 debate, they were determined to join the Buddhists of Ceylon in their battle against Christian missionaries. They sailed to India, arriving in Bombay in 1879, proceeding to Ceylon the next year. There they both took the vows of lay Buddhists in a public ceremony. Olcott wrote in his diary, "We had previously declared ourselves Buddhists long before, in America, both privately and publicly, so that this was but a formal confirmation of our previous professions."[9] Blavatsky's interest in Buddhism remained peripheral to her Theosophy, but Olcott would enthusiastically embrace his new faith, being careful to note that he was a "regular Buddhist" rather than a "debased modern Buddhist sectarian." In a statement that would resonate throughout the subsequent discourse of Buddhism and Science, he wrote:

> Speaking for her as well as for myself, I can say that if Buddhism contained a single dogma that we were compelled to accept, we would not have taken the *pānsil* [the five vows of a lay Buddhist] nor remained Buddhists ten minutes. Our Buddhism was that of the Master-Adept Gautama Buddha, which was identically the Wisdom Religion of the Aryan Upanishads, and the soul of the ancient world-faiths. Our Buddhism was, in a word, a philosophy, not a creed.[10]

Olcott took it as his task to restore true Buddhism to Ceylon and to counter the efforts of the Christian missionaries on the island. In order to accomplish this aim, he adopted many of their techniques, founding the Buddhist Theosophical Society to disseminate Buddhist knowledge (and later assisted in the founding of the Young Men's Buddhist Association) and publishing in 1881 *A Buddhist Catechism*, modeled on works used by the Christian missionaries. Olcott shared the view of many enthusiasts in Victorian Europe and America, w he Buddha as the greatest philosopher of India's Aryan past. The a's teachings were regarded as a complete philosophical and ps ical system, based on reason and restraint, opposed to ritual, su on, and sacerdotalism, demonstrating how the individual could moral life without the trappings of institutional religion. This B was to be found in texts rather than in the lives of the moder uddhists of Sri Lanka, who, in Olcott's view, had deviated from ginal teachings.

His *Buddhist Catechism* contains a chapter entitled "Buddh Science." There we find the following questions and answers like the rest of the book, were meant to be memorized by Sin children):

> **325** Q *Has Buddhism any right to be considered a scientific religion, or may it be classified as a "revealed" one?*
>
> A Most emphatically it is not a revealed religion. The Buddha did not so preach, nor is it so understood. On the contrary, he gave it out as the statement of eternal truths, which his predecessors had taught like himself.[11]

Moving on to discuss the light that Buddhist scriptures describe as emanating from the Buddha's body, he continues:

> **344** Q *What do Europeans call it now?*
>
> A The Human Aura.
>
> **345** Q *What great scientist has proved the existence of this aura by carefully conducted experiments?*
>
> A The Baron von Reichenbach. His experiments are fully described in his Researches—published in 1844–5. Dr. Baraduc of Paris has, quite recently, photographed this light.
>
> **346** Q *Is this bright aura a miracle or a natural phenomenon?*
>
> A Natural. It has been proved that not only all human beings, but animals, trees, plants and even stones have it.[12]

Turning to the Buddha's ability to create illusions, Olcott writes, "*Q. Is this branch of science well known in our day?* A. Very well known; it is familiar to all students of mesmerism and hypnotism."[13]

In Sri Lanka alone, over a period of just a few years, the referents of the terms *Buddhism* and *Science* had shifted. As we shall see in chapter 1, for Guṇānanda, the monk who defended Buddhism in the debate of 1873, Buddhism was the teaching of no self and causation, and Science was a heliocentric cosmology ascribed to Newton and rejected as false. For Olcott, the Buddha was a more magical master whose wisdom also allowed him to manipulate the natural world. Such abilities could be confirmed by mesmerism, an eighteenth-century science devised by Franz Mesmer (the root of the verb *mesmerize*) that purported to use the magnetism present in living species to cure various maladies.

In 1886 Edwin Arnold, author of the best-selling biography of the Buddha, *The Light of Asia*, visited Bodh Gayā, the site of the Buddha's enlightenment in northern India. Finding it in a state of decay and under the control of a Hindu priest, he wrote an essay in the London *Daily Telegraph* decrying its condition. Three years later he visited Japan and was invited to speak at the Imperial University in Tokyo. He concluded his lecture, entitled "The Range of Modern Knowledge" and delivered on December 15, 1889, with the following declaration, one which suggests how encompassing the view of the compatibility of Buddhism and Science had become in the late Victorian period:

> I have often said, and I shall say again and again, that between Buddhism and modern science there exists a close intellectual bond. When Tyndall tells us of sounds we cannot hear; and Norman Lockyer of colours we cannot see; when Sir William Thomson and Professor Sylvester push mathematical investigation to regions almost beyond the Calculus, and others, still bolder, imagine and try to grapple with, though they cannot actually grasp, a space of four dimensions, what is all this except the Buddhist *Maya*, a practical recognition of the illusions of the senses? And when Darwin shows us life passing onward and upward through a series of constantly improving forms towards the Better and the Best, each individual starting in new existence with the records of bygone good and evil stamped deep and ineffaceably from the old ones, what is this again but the Buddhist doctrine of *Dharma* and of *Karma?* And when the Victorian poet and preacher and moralist rightly discern and worthily teach, as the last and truest wisdom, that Justice, Duty, and Right control events, and that the eternal Equity and Compassion of the universe overlooks and forgives no wrong and no disobedience, but also neglects and forgets

> no good deed or word or thought, what is this except the teaching of the Buddha? Finally, if we gather up all the results of modern research, and look away from the best literature to the largest discovery in physics and the latest word in biology, what is the conclusion—the high and joyous conclusion—forced upon the mind, except that which renders true Buddhism so glad and hopeful? Surely it is that the Descent of man from low beginnings implies his Ascent to supreme and glorious developments; that "the Conservation of Matter and Energy," a fact absolutely demonstrated, points to the kindred fact of the conservation and continuity of all Life, whereof matter is but the apparent vehicle and expression; that death is probably nothing but a passage and a promotion; that the destiny of man has been, and must be, and will be worked out by himself under eternal and benign laws which never vary and never mislead; and that for every living creature the path thus lies open, by compliance, by effort, by insight, by aspiration, by goodwill, by right action, and by loving service, to that which Buddhists term Nirvana, and we Christians "the peace of God that passeth all understanding."[14]

Authority in Buddhism is often a matter of lineage, traced backward in time from student to teacher, ideally ending with the Buddha himself. If one were to imagine a lineage of the linkage of Buddhism to Science, one might begin with Guṇānanda (who clearly saw himself as representing the original teachings of the Buddha) to Colonel Olcott, to a young Sinhalese named David Hewaviratne, better known as Anagārika Dharmapāla (1864–1933). He was born into the small English-speaking middle class of Colombo in 1864. His family was Buddhist, but he was educated in Catholic and Anglican schools. He met Blavatsky and Olcott during their first visit to Sri Lanka in 1880 and was initiated into the Theosophical Society four years later, changing his name to Anagārika Dharmapāla ("Homeless Protector of the Dharma"). Although he remained a layman until late in his life, he wore the robes of a monk. He became Colonel Olcott's closest associate, accompanying him on a trip to Japan in 1889.

In 1891, inspired by Edwin Arnold's account of the sad state of the temple of Bodh Gayā, Dharmapāla helped to found the Maha Bodhi Society, whose aim was to wrest the site from Hindu control and make it a place of pilgrimage for Buddhists from around the world, a goal that was not achieved until after his death. He gained international fame after his bravura performance at the World's Parliament of Religions, held in conjunction with the Columbian Exposition in Chicago in 1893. With his eloquent English and ability to quote from the Bible, he captivated audiences as he argued that Buddhism was clearly the equal, if not the superior, of Christianity in both antiquity and profundity, noting, for example, its compatibility with Science. While in Chicago, he met not only the other Buddhist delegates to the parliament, such as the Japanese Zen priest Shaku Sōen, but American enthusiasts of Buddhism, including Paul Carus, a German émigré living in LaSalle, Illinois, and the proponent of what he called "The Religion of Science."[15]

Dharmapāla lectured widely in the years after the Chicago parliament, making frequent reference to the compatibility of Buddhism and Science. At a lecture delivered at the Town Hall in New York in 1925, he declared:

> The Message of the Buddha that I have to bring to you is free from theology, priestcraft, rituals, ceremonies, dogmas, heavens, hells and other theological shibboleths. The Buddha taught to the civilized Aryans of India 25 centuries ago a scientific religion containing the highest individualistic altruistic ethics, a philosophy of life built on psychological mysticism and a cosmogony which is in harmony with geology, astronomy, radioactivity and relativity. No creator god can create an ever-changing, ever-existing cosmos. Countless billions of aeons ago the earth was existing but undergoing change, and there are billions of solar systems that had existed and exist and shall exist.[16]

In an essay published the previous year, "Buddhism, Science and Christianity," the implications of his use of the term *Aryan* are more clear; the "Science" of his title includes one of the most influential sciences of the early twentieth century, race science.

> Every new discovery in the domain of Science helps us to appreciate the sublime teachings of the Buddha Gautama. What is greatly deplored is the attitude of certain European Oriental Scholars who condemn Buddhism without serious study.
>
> The semitic religions have neither psychology nor a scientific back ground. Judaism was an exclusive religion intended only for the Hebrews. It is a materialistic monotheism with Jehovah as the architect of the limited

> world. Christianity is a political camouflage. Its three aspects are the Bible, barrels of whiskey and bullets. It is a religion of ethical contradictions. The old war god of the Jews is yoked with the camouflaged god of love. Whose characteristics are that of a veritable autocrat sending countless millions of people to an eternal hell of fire and brim stone. He enriches the Kingdom of his enemy Satan by increasing the population of hell by a thousand fold. Jesus is camouflaged as the prince of peace, whilst his actions show him to be a personality with an irritable temper. His very disciples forsook him at the critical moment when he prayed for help. He died praying to his god confessing his ignominious failure.[17]

Dharmapāla continues in this vein for several pages, without further mention of science, before concluding: "We condemn Christianity as a system utterly unsuited to the gentle spirit of the Aryan race."[18] The implication of Buddhism in the science of race is discussed in detail in chapter 2.

In 1893, on his journey back from the World's Parliament of Religions, Dharmapāla stopped in Shanghai, where he met Yang Wenhui (1837–1911). Yang was a civil engineer who had become interested in Buddhism after happening upon a copy of *The Awakening of Faith*, one of the more influential works of the huge Chinese Buddhist canon. He organized a lay society to spread the teaching of the Buddha by carving woodblocks for the printing of the Buddhist canon. After serving as a diplomat at the Chinese embassy in London, he resigned from his government position to devote all his energies to the publication of Buddhist scriptures.

When Dharmapāla visited China, he was accompanied by another participant of the World's Parliament, the Reverend Timothy Richard (1845–1919), a well-known Welsh Baptist missionary. Upon his arrival, Dharmapāla tried, unsuccessfully, to enlist Chinese monks into the Maha Bodhi Society. Rev. Richard, who had been investigated by the Baptist Missionary Society in England for teaching something other than "essential theology," arranged for the Sinhalese Buddhist reformer to meet Yang Wenhui. Despite Dharmapāla's urging, Yang did not feel that Chinese monks could be persuaded to travel to India to help restore Buddhism there; he suggested instead that Indians be sent to China to study the Buddhist canon. We observe here one of the issues that beset Buddhist leaders in the late nineteenth century and that has persisted throughout the history of Buddhism and Science: they often did not agree on what constitutes true Buddhism. Dharmapāla, like the leading orientalists of the day, believed that the Pāli scriptures of his native Sri Lanka were the most pure and authentic version of the Buddha's teachings. He would have regarded most of the works that Yang was publishing as inauthentic. Yang, as a proponent of the Mahāyāna, believed that the Buddhism of China was the most complete and authentic. His plan for restoring Buddhism in India was to return the Chinese canon of translated Indian texts (including the Mahāyāna sūtras) to the land of their origin.

Yang and Dharmapāla corresponded over the next fifteen years, agreeing on the importance of spreading Buddhism to the West. Toward that end, Yang collaborated with Rev. Richard in an English translation of *The Awakening of Faith*, published in 1907. In the following year, Yang established a school to train Buddhist monks to serve as foreign missionaries. His contact with figures such as F. Max Müller (whom he had met during his time in England) and Dharmapāla had convinced him that Buddhism was the religion most suitable for the modern scientific world.

The dilemma faced by Buddhists of China was different from that faced by the Buddhists of Sri Lanka. The challenge came not only from Christian missionaries, but from a growing community of Chinese intellectuals who saw Buddhism as one of several forms of primitive superstition preventing China's entry into the modern world; Buddhist practice was fraught with ghosts and demons, and Buddhist doctrine was "life denying." As a religion imported from abroad, Buddhism had periodically been regarded with suspicion by the state over the course of Chinese history and had been subjected to imperial persecution on four occasions (in 564, 567, 845, and 955 CE). Criticism of Buddhism intensified in the early decades of the twentieth century (especially after the Republican revolution of 1911) when it was denounced both by Christian missionaries and by Chinese students who returned from studying abroad, among whom the works of John Dewey and Karl Marx were particularly popular. In 1898 the Qing emperor issued an edict ordering that many Buddhist temples (and their often substantial landholdings) be converted to public schools. Although the order was rescinded in 1905, Buddhist leaders saw this as an opportunity to respond to criticism (and prevent the seizure of their lands), converting monasteries to schools and academies for the training of monks. Modeled to a certain degree on Christian seminaries, the monastic schools set out to train monks in the Buddhist classics. The monks would in turn go out in public and teach these texts to the laity. Yang's academy was one of a number of such schools. Although most closed after a few years, they trained many of the future leaders of modern Buddhism in China, who defended the dharma through founding Buddhist organizations, publishing Buddhist periodicals, and leading lay movements to support the saṅgha (monastic community).[19] One of the students at Yang's school was the monk Taixu.

Taixu (1890–1947) would become one of the most famous figures of the Chinese Buddhist revival of the first decades of the twentieth century. He was ordained as a monk at the age of fourteen, reportedly because he wanted to acquire the supernormal powers of the Buddha. He studied under the tutelage of the famous Chinese monk Eight Fingers (so named because he had burned off one finger of each hand as an offering to the Buddha) and had experienced awakening when reading a perfection of wisdom sūtra. After an early flirtation with revolutionary politics, in 1911 he organized the Association for the Advancement of Buddhism; in 1912 he was involved in a failed attempt to turn the great Buddhist monastery of Chin Shan into a modern school for monks. Disgraced by this failure, he went into retreat for three years, beginning in 1914, during which time he studied the Buddhist scriptures and formulated plans for the reorganization of the community of Buddhist monks. These called for improved and modernized education for monks and their participation in community and government affairs. Taixu would design various versions of such plans over the course of his career; none were adopted. He was also involved in the publication of a wide variety of Buddhist periodicals, such as *Masses Enlightenment Weekly*, *Voice of Enlightenment*, *Buddhist Critic*, *New Buddhist Youth*, *Modern Sangha*, *Mind's Light*, and the most enduring, *The Voice of the Sea Tide*. In 1923 he founded the first of several of what he called "world Buddhist organizations." He began to travel widely, becoming well known in Europe and America.[20] A collection of his essays published in 1928 contains a biographical sketch, where he is described as "the most eminent teacher of Buddhism in recent years"; it concludes, "On August 11 last [1928], he sailed for Europe in order to carry the light of Buddhism to the West, and impart to all a supreme, universal, and absolute perception of the Cosmos."[21]

In his essays in these periodicals as well as in his lectures in China and abroad, Taixu spoke often of the relation between Buddhism and Science. The following is from a lecture published in 1928:

> Scientific knowledge can prove and postulate the Buddhist doctrine, but it cannot ascertain the realities of the Buddhist doctrine. Scientific discoveries have brought about a certain doubt as to religious evidence. The old gods and religions seem to have been shaken in the wind of science, and religious doctrines have no longer any defence, and the world at large seems to be handed over to the tyranny of the machine and all those monstrous powers to which Science has given birth.
>
> Buddhism takes quite a different view, and holds that Science does not go far enough into the mysteries of Nature, and that if she went further the Buddhist doctrine would be even more evident. The truths contained in the Buddhist doctrine concerning the real nature of the Universe would greatly help Science and tend to bring about a union between Science and Buddhism. . . .
>
> By the use of such scientific methods the Buddhist scholar is aided in his research. When we go beyond these methods we find that Science is unable to grasp the reality of the Buddhist doctrine. The reality of the Buddhist doctrine is only to be grasped by those who are in the sphere of supreme and universal perception, in which they can behold the true nature of the Universe, but for this they must have attained the wisdom of Buddha himself, and it is not by the use of science or logic that we can expect to acquire such wisdom. Science therefore is only a stepping stone in such matters.[22]

We see here one of the more strident views of the relation of Buddhism and Science, a view that would be repeated in the future: that Science can confirm the insights of the Buddha but is incapable of gaining those insights through its own means.[23] Science—identified here, as is often the case in the Buddhism and Science discourse, with a somewhat bewildering and frightening technology—can be of benefit in investigating the world, but its scope is limited. In order to penetrate beyond appearances to the true nature of reality, however, science is inadequate. To understand the nature of reality, one must achieve buddhahood.

In the last decades of the nineteenth century, Buddhism was also under attack in Japan. In an effort to demonstrate the relevance of Buddhism to the larger interests of the Japanese nation, Buddhist leaders sought to promote a New Buddhism that would play an active role in Japan's attempts to modernize and expand. In the early years of the Meiji period (1868–1912), Buddhism had been attacked in terms that would be repeated in China at the end of the nineteenth century. It was portrayed as a foreign and anachronistic institution, riddled with corruption, a parasite on society, and the purveyor of superstition, impeding Japan from taking its rightful place among the great nations of the modern world. The New Buddhism that was espoused in response to such charges was represented as both purely Japanese and purely Buddhist—more Buddhist, in fact, than the other Buddhisms of Asia, especially those of China and Korea, which many Japanese Buddhist leaders conceded were corrupt. The New Buddhism was also committed to social welfare, urging the foundation of public education, hospitals, and charities. It supported the military expansion of the Japanese empire. And it was fully consistent with modern science.[24]

One of the leading figures of the New Buddhism was Shaku Sōen (1859–1919). Ordained at the age of twelve as a novice in the Rinzai sect of Zen, he studied under the master Imakita Kosen (1816–1892). He trained under Kosen at the famous Engakuji monastery in Kamakura, receiving "dharma transmission," and hence authority to teach, at the age of twenty-four. Seeking to combine both Buddhist training and Western-style education, he attended Keio University and then traveled to Ceylon to study Pāli and live as a Theravāda monk; by this time, the European orientalist view had reached Japan that held that the Pāli scriptures of Theravāda Buddhism were closest to the original teachings of the Buddha. Upon his return, Shaku Sōen was chosen by a conference of abbots to be one of the four editors of a work to be entitled *The Essentials of All the Buddhist Sects*. He was selected to be one of the Japanese representatives to the World's Parliament of Religions in 1893; his address was translated into English by one of his lay disciples, D. T. Suzuki. In his address to the parliament, Shaku Sōen did not speak about Bodhidharma coming to the West, the koan *mu*, the practice of zazen, or the experience of *satori*. He spoke about causation.

> If we open our eyes and look at the universe we observe the sun and moon and stars in the sky; mountains, rivers, plants, animals, fishes and birds on the earth. Cold and warmth come alternately; shine and rain change

> from time to time without ever reaching an end. Again let us close our eyes and calmly reflect upon ourselves. From morning to evening we are agitated by the feelings of pleasure and pain, love and hate; sometimes full of ambition and desire, sometimes called to the utmost excitement or reason and will. Thus the action of mind is like an endless issue of a spring of water. As the phenomena of the external world are various and marvelous, so is the internal attitude of the human mind. Shall we ask for the explanation of these marvelous phenomena? Why is the universe in constant flux? Why do things change? Why is the mind subjected to a constant agitation? For these Buddhism offers only one explanation, namely the law of cause and effect.[25]

Thus, among all the topics that Shaku Sōen, a Zen priest and Mahāyāna master, could have chosen to introduce Buddhism to his American audience, he chose not emptiness or compassion or the Buddha nature, but the comparatively prosaic topic of causation. It is in fact a basic Buddhist doctrine accepted, at least on the conventional level, by all schools of Buddhism across Asia. But it is likely that Shaku Sōen also selected it because it seemed utterly modern, and scientific, explaining both the outer world of matter and the inner world of mind without recourse to God.[26]

Indeed, it was in the late nineteenth century that Buddhist apologists began to refer to the doctrine of karma as a natural law. In classical Buddhist doctrine, all human experiences of pleasure and pain are the result of deeds done in the past, either in the present lifetime or in one of innumerable past lives. Evil or nonvirtuous (*akuśala*) deeds of body, speech, or mind—that is, physical actions, words, and thoughts—inevitably result in physical or mental suffering in the future, either in the present lifetime or some future life, unless the seeds of suffering are destroyed by wisdom. These deeds are classically enumerated as ten: killing, stealing, sexual misconduct, lying, divisive speech, harsh speech, senseless speech, covetousness, harmful intent, and wrong view. Virtuous deeds of body, speech, and mind inevitably result in physical or mental happiness in the future. It might be said that just as an object dropped from a table will fall to the ground by the force of gravity, a virtuous deed will result in a feeling of happiness by the force of karma. Thus, when the model of a mechanistic universe prevailed in European science, Buddhists could claim that such a mechanism also pertained in the realm of morality, that according to the natural law of karma, the universe was ethical, without the need for God to bestow blessings or mete out punishments. Describing the law of karma, Thomas Huxley, the great proponent of Darwin, wrote in 1894, a year after the Chicago parliament where Shaku Sōen had spoken:

> If this world is full of pain and sorrow; if grief and evil fall, like the rain, upon both the just and the unjust; it is because, like the rain, they are links in the endless chain of natural causation by which past, present, and future are indissolubly connected; and there is no more injustice in the one case than in the other. Every sentient being is reaping as it has sown; if not in this life, then in one or other of the infinite series of antecedent existences of which it is the latest term. The present distribution of good and evil is, therefore, the algebraical sum of accumulated positive and negative deserts; or, rather, it depends on the floating balance of the account.[27]

Upon his return to Japan, Shaku Sōen reported on the work of the Japanese delegation at the parliament:

> We invited the attention of the participants, both foreign and Japanese, to the following points at least: that the Japanese are people with abundantly loyal and patriotic spirits; that Buddhism has exercised great influence on Japanese spirituality, and had influence on successive emperors too; that Buddhism is a universal religion and it closely corresponds to what science and philosophy say today; that we cleared off the prejudice that Mahāyāna Buddhism was not the true teaching of the Buddha; that Mr. Straw, a wealthy merchant in New York, had a conversion ceremony carried on at the congress hall in which he became a Buddhist; that a leading Japanese staying in the United States arranged a Buddhist lecture meeting twice for us in the Exposition building, and so on.[28]

This passage summarizes the multiple motivations of the Japanese delegation to the Chicago parliament. In addition to seeking to demonstrate the historical links between Buddhism and the Japanese imperial family as well as the relevance of Buddhism to the modern world (and hence its compatibility with Science), they were concerned to establish the legitimacy of the Mahāyāna (which they also referred to as Eastern Buddhism) as an authentic form of Buddhism, against the claims of both Theravāda delegates and the scholarly opinion of the day that the Pāli scriptures of the Theravāda most faithfully represented the Buddhism of the Buddha. Although not alluded to in this passage, the Japanese delegation also sought to draw attention to the unfavorable terms of Japan's treaties with the United States; the other lecture that Shaku Sōen delivered at the parliament was entitled "Arbitration Instead of War."[29] He later served as a chaplain in Manchuria during the Russo-Japanese War of 1904–5; by that time he had become sufficiently well known that Leo Tolstoy sought to enlist his support in condemning the war between their two nations, a request that the Zen priest refused.

When Shaku Sōen's disciple D. T. Suzuki came to compose his first major work in English, he wrote not about the Zen that would make him so famous. Instead he continued the work that Shaku Sōen had begun at the World's Parliament of Religions. Thus, in his 1907 *Outlines of Mahayana Buddhism*, Suzuki describes Buddhism as a universal religion, he defends the Mahāyāna as the genuine teaching of the Buddha, and he sets forth its correspondence to Science. Indeed, as others had argued in the past and would argue in the future, Suzuki announced that the discoveries of science had been anticipated by the Buddha. He wrote, "It is wonderful that Buddhism clearly anticipated the outcome of modern psychological researches at the time when all other religious and philosophical systems were eagerly cherishing dogmatic superstitions concerning the nature of the ego. The refusal of modern psychology to have soul mean anything more than the sum-total of all mental experiences, such as sensations, ideas, feelings, decisions, etc., is precisely a rehearsal of the Buddhist doctrine of non-âtman."[30] At the same time, he declared the compatibility of Buddhism and Science.

> It is to be infered [sic] that Buddhism never discourages the scientific, critical investigation of religious beliefs. For it is one of the functions of science that it should purify the contents of a belief and that it should point out in which direction our final spiritual truth and consolation have to be sought. Science alone which is built on relative knowledge is not able to satisfy all our religious cravings, but it is certainly able to direct us to the path of enlightenment. When this path is at last revealed, we shall know how to avail ourselves of the discovery, as then Prajñâ (or Sambodhi, or Wisdom) becomes the guide to life. Here we enter into the region

of the unknowable. The spiritual facts we experience are not demonstrable, for they are so direct and immediate that the uninitiated are altogether at a loss to get a glimpse of them.[31]

We have now surveyed the works of Buddhist leaders in Sri Lanka, China, and Japan during the period before the First World War, and the messages that they delivered to audiences in their homelands and in Europe and America. Although each offered a somewhat different view from the other, together they regard Buddhism as that religion most compatible with Science, although they also hold that Buddhism offers access to states of wisdom that Science alone can never attain. Some even went so far as to declare that Buddhism was not a religion at all, but was itself a science of the mind. The implications of such a claim become clear in light of theories of social evolution of the day, which saw an inevitable advance of humanity from the state of primitive superstition to religion to science. By claiming it to be a science, Buddhism, condemned as a primitive superstition both by European missionaries and by Asian modernists, leaps from the bottom of the evolutionary scale to the top.

Although it has become fashionable in academic circles to speak not of a single Buddhism but of many Buddhisms (located at different points in history and different sites in the world), in this case it seems possible, despite differences of language and of cultural context, to ascribe a certain coherence to the discourse of Buddhism and Science. The passages cited above were all written within a relatively brief period of time. And the authors of these passages were, in one way or another, related to each other. Olcott was inspired to visit Sri Lanka by the debate of Guṇānanda, whom Olcott met. Olcott then worked closely with Dharmapāla, who in turn met Yang Wenhui, who was a teacher of Taixu. At the Chicago parliament, Dharmapāla met Shaku Sōen, and they both met Paul Carus. Shaku Sōen would later arrange for his student D. T. Suzuki to come to America, where he lived and worked with Carus for eleven years.

This is not to suggest that the lineage went unchallenged or that the claims of Buddhism and Science were universally affirmed. Voices of dissent were occasionally heard, some emanating from the academy. In 1923 the British Indologist and jurist Arthur Berriedale Keith wrote:

> Yet another, and perhaps more serious, defect in the most popular of current expositions of Buddhism is the determination to modernize, to show that early in Buddhist thought we had fully appreciated ideas which have only slowly and laboriously been elaborated in Europe, and are normally regarded as the particular achievement of modern philosophy. Now there is nothing more interesting or legitimate than, on the basis of a careful investigation of any ancient philosophy, to mark in what measure it attains conceptions familiar in modern thought; but it is a very different thing to distort early ideas in order to bring them up to date, and the futility of the process may be realized when it is remembered that every generation which yields to the temptation will succeed in finding its own conceptions foreshadowed.[32]

The discourse of Buddhism and Science remained relatively dormant during the 1940s and 1950s, with the exception of parallels drawn by some between psychoanalysis and Zen, inspired largely by the works of D. T. Suzuki.[33] It reemerged in the 1960s with the efflorescence of interest in Asian religions and Eastern wisdom. The signal publication during this period was the improbable best-seller, *The Tao of Physics*, first published in 1975. The work went on to become a classic of the New Age, selling more than 1 million copies.

Its author, Fritjof Capra (b. 1939), was a physicist rather than a student of Buddhism, and thus relied on secondary sources for his portrayal of Buddhist thought. As was the case with a number of works of this period, Capra spoke as often of "Eastern mysticism" as of a specific tradition. This mysticism included, then, insights of Hinduism, Buddhism, and Taoism.

> When I refer to "Eastern mysticism," I mean the religious philosophies of Hinduism, Buddhism, and Taoism. Although these comprise a vast number of subtly interwoven spiritual disciplines and philosophical systems, the basic features of their world view are the same. The view is not limited to the East, but can be found to some degree in all mystically oriented philosophies. The argument of this book could therefore be phrased more generally, by saying that modern physics leads us to a view of the world which is very similar to the views held by mystics of all ages and traditions. Mystical traditions are present in all religions, and mystical

> elements can be found in many schools of Western philosophy. The parallels to modern physics appear not only in the *Vedas* of Hinduism, in the *I Ching*, or in the Buddhist *sutras*, but also in the fragments of Heraclitus, in the Sufism of Ibn Arabi, or in the teachings of the Yaqui sorcerer Don Juan.[34]

Like the Theosophists of the nineteenth century, Capra sees a deep foundation from which all mystical traditions arise, a tradition that both anticipates and is confirmed by what he calls "the New Physics." Both mysticism and modern physics derive their insights from empirical methods, yet those insights cannot be expressed in words. They thus remain beyond the comprehension of those who are neither mystics nor physicists. As he writes, "A page from a journal of modern experimental physics will be as mysterious to the uninitiated as a Tibetan mandala. Both are enquiries into the nature of the universe."[35] For the mystic, the result of his inquiry is "a direct non-intellectual experience of reality."

> A Hindu and a Taoist may stress different aspects of the experience; a Japanese Buddhist may interpret his or her experience in terms which are very different from those used by an Indian Buddhist; but the basic elements of the world view which has been developed in all these traditions are the same. These elements also seem to be the fundamental features of the world view emerging from modern physics.
>
> The most important characteristic of the Eastern world view—one could almost say the essence of it—is the awareness of the unity and mutual interrelation of all things and events, the experience of all phenomena in the world as manifestations of a basic oneness. All things are seen as interdependent and inseparable parts of this cosmic whole; as different manifestations of the same ultimate reality.[36]

Capra devotes individual chapters to five forms of Eastern mysticism: Hinduism, Buddhism, Chinese Thought, Taoism, and Zen, implying perhaps that Taoism is not a form of Chinese Thought and that Zen is not a form of Buddhism. In discussing Buddhism, he explains that while Hinduism is "mythological and ritualistic," Buddhism is "psychological." The Buddha's "doctrine, therefore, was not one of metaphysics, but one of psychotherapy."[37] After surveying the four noble truths, he turns to the Mahāyāna, and a text entitled *The Awakening of Faith*, which he describes as a first-century CE work that is "the first representative treatise on the Mahayana doctrine and has become a principal authority for all schools of Mahayana Buddhism."[38] The *Awakening of Mahāyāna Faith* (*Dasheng qixin lun*) is a work attributed to the Indian master Aśvaghoṣa, first translated into Chinese in 553. Japanese scholars have demonstrated that it is in fact an apocryphon, not translated but instead originally composed in China in the sixth century and attributed to Aśvaghoṣa in order to enhance its authority. Its influence in East Asian Buddhism has indeed been profound. But since it is neither a first-century work nor an Indian work, it cannot be "a principal authority for all schools of Mahayana Buddhism." Perhaps more important than its pedigree is the fact that it was one of the first Chinese Buddhist works to be translated into English, and by figures who hold a central place in the discourse of Buddhism and Science. It was translated by D. T. Suzuki and Paul Carus in 1900 and, as noted above, by Timothy Richard and Yang Wenhui in 1907.[39] Indeed, it is Suzuki's influential vision of Zen as an experience, separate from historical contingencies and separate ultimately from Buddhism, that informs much of Capra's work.[40]

After describing the various mystical traditions of Asia, Capra surveys "the New Physics," noting similarities and parallels along the way. His goal, however, does not seem to be one of proof, but of evocation.

> The Eastern religious philosophies are concerned with timeless mystical knowledge which lies beyond reasoning and cannot be adequately expressed in words. The relation of this knowledge to modern physics is but one of its many aspects and, like all the others, it cannot be demonstrated conclusively but has to be experienced in a direct intuitive way. What I hope to have achieved, to some extent, therefore, is not a rigorous demonstration, but rather to have given the reader an opportunity to relive, every now and then, an experience which has become for me a continuing joy and inspiration; that the principal theories and models of modern physics lead to a view of the world which is internally consistent and in perfect harmony with the views of Eastern mysticism.[41]

Capra thus is making a declaration of faith, and inviting others to share in that faith for the joy and comfort it provides. He does not explain the nature of this joy or why his experience is joyful. One can only assume that he finds a deep comfort in the knowledge that what is newly known was once known long ago. It is notable, however, that his own knowledge of the ancient mystics, at least in the case of Buddhism, is drawn not from ancient texts but from modern apologists, most important, D. T. Suzuki, disciple of Shaku Sōen, disciple of Paul Carus, and a founding figure in the discourse of Buddhism and Science.

It was Suzuki's work that was most widely read in the 1950s and 1960s, before Japanese Zen teachers and Tibetan lamas came to America. From reading the works of Suzuki, parallels between Buddhism and Science are easy to discern; indeed, as we have seen, Suzuki draws them himself. When confronted with Asian teachers, the situation is somewhat more complicated, as Capra suggests in his afterword to the third edition of *The Tao of Physics*: "I no longer believe that we can adopt Eastern spiritual traditions in the West without changing them in many important ways to adapt them to our culture. My belief has been enforced by my encounters with many Eastern spiritual teachers who have been unable to understand some crucial aspects of the new paradigm that is now emerging in the West."[42] Perhaps Capra had not met the Dalai Lama.

The Fourteenth Dalai Lama of Tibet has been the most visible and influential Buddhist teacher to embrace the discourse of Buddhism and Science. The figures of the late nineteenth and early twentieth centuries, such as Dharmapāla, Taixu, and D. T. Suzuki, had complicated (and in some cases strained) relations with the Buddhist institutions of their homelands, and none had either the authority or the stature of the Dalai Lama. Born in 1935 in a distant corner of the Tibetan cultural domain, he was identified as a young boy as the fourteenth incarnation of the Dalai Lama, a lineage of Buddhist teachers that extends back to the fifteenth century; since 1642, the Dalai Lamas were also the temporal rulers of Tibet. He displayed an interest in mechanical things from the time of his tutelage in the Potala Palace in Lhasa, where he discovered various European gadgets given as gifts to his predecessor. The People's Liberation Army invaded Tibet in 1950 and the Dalai Lama assumed the position of head of state, traveling in 1954 to China, where he was impressed by Chinese feats of engineering. During a failed Tibetan uprising against the occupying army in March 1959, he escaped to India, where he has lived since.

The Dalai Lama's first decade in exile was dominated by the plight of the tens of thousands of Tibetans who escaped across the Himalayas into India and Nepal. He wrote little during this time, and one finds few references to science. His first book on Buddhism, published in Tibetan in 1963, begins:

> At this time of the twentieth century, an era of chemicals and weaponry, during the phase of ethics among the ten periods of five hundred years in the teaching of the fourth leader, the Teacher [Śākyamuni Buddha], external material culture is continuing to develop and expand. At the same time, there is a vital need for similar development and expansion of inner awareness and attitude.[43]

Writing in Tibetan for a Tibetan audience, the Dalai Lama measures time in a traditional way. Among the many prophecies concerning how long Buddhism will last after the Buddha's passage into nirvāṇa, there is one that predicts that it will last for ten periods, each five hundred years in length, and each associated with a different virtue. According to some systems of reckoning, we now live in the sixth such period, where the chief virtue is ethics. After the entire period of five thousand years has come to an end, the teachings of the Buddha will disappear from the world and be forgotten. At some point in the future, another buddha will appear in the world. Indeed, buddhas have already appeared in our world on three occasions, and their teachings have been forgotten. Hence, the Buddha who appeared in India some twenty-five hundred years ago, referred to by the Dalai Lama simply as "the Teacher," is the fourth buddha to appear in our universe. The Dalai Lama thus begins his book from the perspective of the traditional Buddhist cosmology. He would come to call certain elements of that cosmology into question in subsequent years.

The passage cited above offers a somewhat conservative position on the question of Buddhism and Science. The Dalai Lama acknowledges the great power, and danger, of science, identifying the external world as its domain. Although it may dominate there, Buddhism nonetheless has much to offer for inner development. This is one of the standard tropes of the Buddhism and Science discourse, that Science describes the external world and Buddhism describes the internal world.[44] In the subsequent decades, the Dalai Lama's views on this question would change, as discussed in chapter 3.

Although particular forms of Theravāda and Zen meditation continue to be mentioned in the discourse of Buddhism and Science, in the last decade of the twentieth century and the first decade of the twenty-first, the primary referent of the term *Buddhism* in the phrase "Buddhism and Science" has been Tibetan Buddhism. Just a century before, Tibetan Buddhism was regarded as the most corrupt and least authentic form of Buddhism, not only by European scholars but by some Asian Buddhists as well. It was seen to have deviated so far from the original teachings of the Buddha that it did not merit the name Buddhism and was called instead Lamaism. Here is the opening paragraph of the entry on Lamaism from the ninth edition of the *Encyclopædia Britannica* (published between 1875 and 1888) by Thomas W. Rhys Davids, author of the entry on Buddhism cited earlier:

> LĀMĀISM is partly religious, partly political. Religiously it is the corrupt form of Buddhism prevalent in Tibet and Mongolia. It stands in a relationship to primitive Buddhism similar to that in which Roman Catholicism, so long as the temporal power of the pope was still in existence, stood to primitive Christianity. . . . Lāmāism is hardly calculated to attract much attention for its own sake. Tibetan superstitions and Tibetan politics are alike repugnant to Western minds. But, as so many unfounded beliefs and curious customs have a special value of their own to the student of folklore, so Lāmāism has acquired a special interest to the student of comparative history through the instructive parallel which its history presents to that of the Church of Rome.[45]

This view remained sufficiently widespread in the twentieth century that the Dalai Lama felt the need to address it in his 1963 treatise, where he writes, "Some people say that the religion of Tibet is 'Lamaism,' as if it were a religion not taught by the Buddha, but this is not so. The original author of the sūtras and tantras that are the root source of all the schools of Tibetan Buddhism is the teacher Śākyamuni Buddha."[46] It is important to note that Tibetan Buddhism, which was regarded as superstition and folklore at the end of the nineteenth century, became the conversation partner of neurobiology at the end of the twentieth.

Through these various peregrinations, the discourse of Buddhism and Science has survived from the nineteenth, through the twentieth, and now into the twenty-first century. It began in the arena of polemics, with Buddhists seeking to defend their religion against the attacks of Christian missionaries. In its first stage, *Buddhism* referred most often to the

Theravāda Buddhism of Sri Lanka (one that rejects the Mahāyāna sūtras as the word of the Buddha), and *Science* referred for the most part to theories of a heliocentric universe, views that had had their own troubled past in the history of Christianity. In the next phase, *Buddhism* became the Buddhism of Blavatsky, an eccentric blend of orientalist scholarship seen through the lens of Theosophy. This Buddhism was claimed to be an ancient wisdom, of which the various Buddhisms of contemporary Asia were but mere reflections. And this Buddhism was seen as compatible with quack sciences, such as mesmerism and the theory of auras. Asian Buddhist leaders, after embracing Theosophy as an ally, would come to reject it, turning this time to another European source, this one more reputable: the Buddhism of the orientalists. This Buddhism was the philosophy of the Buddha, as they understood him, an aristocratic teacher who rebelled against the corrupt priestcraft of his day to teach an ethical system that required no God, and which opened the path to freedom from suffering to all men. Asian Buddhists and European enthusiasts could thus claim Buddhism as the most modern of the world religions, able to uphold morality without the need for an angry creator God, and as the most scientific, fully in accord with the science of the day, which described a mechanistic universe of cause and effect.

In the period after the Second World War, this science was displaced by Einstein's theories and Theravāda Buddhism was displaced by Zen, especially as set forth by D. T. Suzuki. The focus turned from cause and effect to relativity and from the law of karma to "interdependence," through creative readings of Nāgārjuna's statements on *pratītyasamutpāda*, "dependent origination." In more recent years, expositions of emptiness and quantum physics have continued (although now drawing on Tibetan interpretations of Indian Buddhist doctrine), with a new element added: the relation of Buddhism to cognitive science, especially through laboratory investigations of the effects of Buddhist meditation on the brain.

Thus, over the course of a century and a half, *Buddhism* has meant the Theravāda tradition of late nineteenth-century Sri Lanka, the "esoteric Buddhism" of Theosophy, the ethical Buddhism of the orientalists, the Zen of D. T. Suzuki, the Madhyamaka philosophy of Nāgārjuna, and the Mahāyāna and tantric Buddhism of Tibet. Science has meant basic astronomy, a mechanistic universe, modern physics, modern cosmology, and neurobiology. The referent of *Buddhism* and the referent of *Science* have changed radically over the course of more than a century, yet the claim for the compatibility of Buddhism and Science has continued to be made. And in each case, in order for the claim to be made, each term must be radically restricted. *Buddhism* becomes a single tradition, and within that tradition, an isolated set of elite doctrines and practices. The term *science* is often restricted to such an extent that it is like a mantra, a potent sound with no semantic value.

By the time that the first claims of affinity between Buddhism and Science began to be made in Asia in the late nineteenth century, Science had come to carry connotations of authority, validation, and truth, separate from and, in some cases, in conflict with, those of the Christian church. It is therefore perhaps unsurprising that Buddhist leaders in Asia would point to what they identified as the scientific aspects of Buddhism in an effort to trump the charges of idolatry and superstition leveled at them by Christian missionaries across the Buddhist world. They argued that the Buddha knew long ago what the science of the Christian West was only now discovering, whether it be the mechanisms of causation that rely on no god, the analysis of experiences into their component parts, the subtle disintegration of matter called impermanence, or the existence of multiple universes.

In the discourse of Buddhism and Science, it often seems that *Science* refers to the eternal truths in nature, and in the mind. Some of those truths have been discovered, some await discovery. But Science has also not been immune from the historical, the social, the political.[47] Buddhists first encountered "Science," perhaps ironically, in the guise of Christianity; it was a superior knowledge, a knowledge that Christianity possessed and Buddhism did not, thus providing yet further proof of the superiority of Christianity, and hence a tool of the missionary and a reason for conversion. Later, Science would be portrayed as the product of a more generalized "European civilization," something that this civilization would take around the world; the vehicle for that journey was colonialism.

The modern Buddhists of the late nineteenth and early twentieth centuries thus had good reason to try to claim science for themselves. Whether to counter the missionary's charge that Buddhism was superstition and idolatry, or to counter the colonialist's claim that the Asian was prone to fanciful flights of the mind and meaningless rituals of the body, or to counter both, science proved the ideal weapon. It was Buddhism, in fact, that was the scientific religion, the religion best suited for modernity, throughout the world. It was an Asian, the Buddha, who knew millennia ago what the European was just beginning to discover. This latter point was only made possible through the strange international network that invented the Buddha as we know him (described in chapter 4).

A century later, the missionaries have not gone away, but their inroads into Buddhist societies are largely confined to specific times and places of the past: Japan in the sixteenth century, Sri Lanka in the early nineteenth century, Korea in the late nineteenth century. And European colonialism, in its classical form, has died out. Yet the discourse of Buddhism and Science persists, unchanged in so many ways. The enemy is slain; the weapon continues to be wielded.

The Buddhist figures mentioned above and considered in the following chapters are a disparate group, coming from different Buddhist cultures, regarding different forms of Buddhism as the most authentic. However, if we consider them from another perspective, they have something in common. Guṇānanda and Dharmapāla were patriots in the struggle for Sri Lankan independence, Yang Wenhui and Taixu sought to prove the relevance of Buddhism to the new Republic of China, Shaku Sōen and Suzuki argued for the essential role for Buddhism in the expanding empire of Japan, and the Tibetan writer Gendun Chopel (discussed in chapter 1 and chapter 3) was imprisoned for acts of treason, acts inspired by the conviction that Tibet must be transformed into a modern nation-state. The inevitable links between nation and science help to explain why today the most famous proponent for the links between Buddhism and Science is none other than the Dalai Lama, who has struggled for a half century for the independence of Tibet, perhaps still seeking to demonstrate that Tibetan Buddhism is not the primitive superstition that the European orientalists saw in the nineteenth century and that the Chinese Communist Party saw in the twentieth. Rather, Tibetan Buddhism is presented as a worthy interlocutor of Science and hence an appropriate ideology of a modern nation that might one day exist.

Such an argument may explain the motivations of the Dalai Lama and the other Buddhist leaders who have invoked Science over the decades. But this alone is not enough to sustain the discourse into the twenty-first century.

Buddhism is renowned for its ability to assimilate the traditions of its environment. In India, the Hindu gods were retained, but deprived of their immortality and consigned to the round of rebirth. With the rise of the Mahāyāna four centuries after the passing of the Buddha, the earlier tradition became "the lesser vehicle" (*hīnayāna*), still a path to liberation but inadequate for the greater goal of buddhahood. With the rise of tantra, it was claimed not only that all those who sought buddhahood must enter the tantric path, but that all the buddhas of the past, including Śākyamuni, had also done so. In Japan, the master Kūkai ranked Confucianism second from the bottom in a hierarchy of ten levels of teachings, with his own Shingon sect at the top. In Burma, alchemical practices designed to prolong life were used to allow one to live long enough to meet the future buddha Maitreya. In Tibet, the fierce gods of the mountains and valleys were tamed by Buddhist masters, pledging to forever protect the dharma rather than lose their lives.

This description of the spread of Buddhism falsely ascribes a kind of agency to the abstract entity "Buddhism," moving from culture to culture around the world and absorbing all that it encounters into itself. Perhaps even more falsely, such a description implies that there is a Buddhism that exists prior to and apart from its assimilated elements. Nonetheless, this quality of accommodation is a powerful metaphor in Buddhism, acknowledging and accepting all that is not somehow the true Buddhism as being of real, but only temporary, value, bringing worldly good but unable to effect the ultimate aim. The ultimate aim is the exclusive domain of the dharma, however it might be understood. The Dalai Lama recently said, "I have great respect for science. But scientists, on their own, cannot prove nirvana. Science shows us that there are practices that can make a difference between a happy life and a miserable life. A real understanding of the true nature of the mind can only be gained through meditation."[48] His comments provide a possible model for the relationship of Buddhism and Science: Buddhism and Science are two different domains, "non-overlapping magisteria," in Stephen Jay Gould's memorable phrase. The Buddha taught the ultimate truth, the nature of and path to liberation from rebirth and nirvāṇa. Science is concerned with the conventional truth, the mundane nature of the world, and offers much insight into its operation. Yet each tradition, Buddhism and Science, has produced important tomes on both the conventional and the ultimate. And where the line between the ultimate and the conventional must be drawn is a question that Buddhist thinkers have pondered, and contested, for two millennia.

Buddhism has long enjoyed the position, first in the West and then throughout much of the world, as the religion that is not a religion, as the ideal alternative to theism, dogma, and ritual, as the religion that spread across Asia not through war but through the self-evident goodness of its teachings and the benevolent teacher who taught them. This vision of Buddhism persists in the face of decades of scholarship that has demonstrated that the history of Buddhism is a good deal more complicated, as a single moment of reflection would seem to suggest.

With this special status so firmly in place, would it be possible for Buddhism to renounce its attachment to Science? Or is Buddhism's apparent compatibility with Science, regardless of the difficulties suggested here, now an essential and inevitable element of its mystique? Is it a necessary step in the evolution of Buddhism? This book offers an occasion to ponder such questions.

It is composed of five essays and a conclusion. The first chapter considers the rise and fall of Mount Meru, the great peak that stands at the center of the (flat) Buddhist world. It was an early target of Christian missionaries, who sought to show that Buddhist beliefs were incompatible with the discoveries of European astronomy and geography. Its existence was defended by some Buddhist monks, but Mount Meru would eventually crumble, the first constituent of Buddhist doctrine that was deemed dispensable in order for Buddhism to be compatible with Science. The tale of the sad fate of this central mountain is recounted here.

The second chapter discusses Buddhist attitudes toward social class, and the representation of those attitudes by both European and Buddhist thinkers in the nineteenth century. The Buddha's apparent rejection of caste distinctions was described by Eugène Burnouf as early as 1844 as "this celebrated axiom of Oriental history," and would be repeatedly invoked in portrayals of Buddhism's harmony with the principles of the European Enlightenment. As some of the early orientalists noted, the Buddhist attitude toward caste was not as simple as it was often depicted. Apart from providing some background on the social setting from which Buddhism emerged in India, one might ask what the question of caste has to do with Buddhism and Science. The answer, in brief, is that the Buddhist notion of superiority, expressed by the Sanskrit term *āryan*, would play a role in perhaps the most notorious science of the late nineteenth and early twentieth centuries, the science of race.

The third chapter looks closely at the contributions of two Tibetans to the discourse of Buddhism and Science. In the early decades of the discourse, as noted above, Tibetan Buddhism, called Lamaism, was sometimes not regarded as a legitimate form of Buddhism by European scholars, and was excluded from the conversation. Tibetan Buddhists were indeed latecomers to Buddhist modernism. The most sustained Tibetan discussion of Buddhism and Science in the first half of the twentieth century came from the renegade scholar Gendun Chopel (1903–1951), who in 1938 published an essay arguing that the world is round. His more extensive views on Buddhism and Science had little influence because they were not published until 1990, long after his death. They are translated in their entirety in this chapter. The other Tibetan is of course the Fourteenth Dalai Lama, who over the past two decades has become the most influential Buddhist voice in the discourse of Buddhism and Science. His views on a range of scientific discoveries, and their implications for Buddhist doctrine, are examined here. Victorian enthusiasts often claimed a compatibility of the Buddhist doctrine of karma with Darwin's theory of evolution, arguing that Buddhism was thus somehow above the debates that have raged over evolution and creationism. The Dalai Lama asserts, however, that the theory of natural selection is not something that Buddhism should easily accommodate.

The fourth chapter explores the rise of the academic study of Buddhism in Europe, what I call "the science of Buddhism," especially in its strained relations with perhaps the most influential of the spiritual sciences of the nineteenth century, Theosophy. In order to understand the referent of *Buddhism* in the development of the discourse of Buddhism and Science, it is essential to examine the series of events that caused Buddhism (for the most part, in the form of Buddhist texts) to become an object of scholarly investigation in Europe, and later in America. The Buddha who appears in the discourse of Buddhism and Science over the course of the past century is largely the product of this scholarship, a scholarship that, for most of its history, operated without the active, or at least acknowledged, participation of Asian Buddhists. The Asian figures who participated most fully in the discourse of Buddhism and Science turned not to their own traditions for their view of the Buddha, but to the European science of Buddhist Studies.

The final chapter briefly examines the most current development in the long history of Buddhism and Science, the attempt to assess the validity of Buddhist meditation through neurological research. Unlike much of the discourse on Buddhism and Science over the decades, which has largely been devoted to identifying affinities between particular Buddhist doctrines and particular scientific theories, this recent research on meditation seeks to measure the physiological and neurological effects of Buddhist meditation in a laboratory setting. Although the ubiquity of meditation in Buddhism is often overstated, meditation has long been represented within the Buddhist tradition as the practice par excellence, and it is apparent that the Buddhist monks who became virtuoso meditators over the centuries derived some form of psychological benefit from it. To be able to define what that was would be of considerable interest. Rather than attempting to survey the burgeoning literature on this topic, in this chapter I ask what is entailed in seeking to determine whether Buddhist meditation works. The chapter, and the book, concludes not with answers to the questions raised throughout but rather with an attempt to refine those questions.

In 1905 W. S. Lilly observed that "one secret of the marvellous success of Buddhism—'that Protean creed,' as Bishop Bigandet calls it—is to be found in its power of accommodating itself to the minds and ways of the populations that received it."[49] As noted above, Buddhism made its remarkable migration across Asia through a process of assimilation. In India, the Vedic gods became disciples of the Buddha. There were traditionally thirty-three Vedic gods. The lowest of the Buddhist heavens, a place of blissful but temporary rebirth, is called the Heaven of the Thirty-Three. In Tibet, the fierce protectors of the snowy peaks were defeated in mystic battle by Buddhist masters and, in exchange for their lives, took oaths to always protect Buddhism. In Japan, the more prominent of the local spirits, the *kami*, were identified as manifestations of various bodhisattvas. In order for Buddhism to establish itself in Europe and America, must the god of the West, the god of Science, also find its place in the Buddhist pantheon?

FIRST THERE IS A MOUNTAIN

On the morning of August 26, 1873, five thousand people gathered around a large platform in the town of Pānadurē, outside Colombo in Sri Lanka. The platform, constructed especially for the occasion, was divided into two sections. One side, with a table covered in white cloth and adorned with evergreens, was occupied by a group of Protestant clergymen. The other side was more richly decked, with tablecloths of damask and a canopy of red, white, and blue cloth. It was occupied by two hundred Buddhist monks in saffron robes. A debate would be held over the next two days. A reporter from the *Ceylon Times* described the scene:

> The time appointed for commencing the discussion was eight o'clock in the morning, and long before that hour, thousands of natives were seen wending their way, attired in their gayest holiday suits, into the large enclosure in which stood the ample bungalow where the adversaries were to meet. By seven the green was one sea of heads. . . . Larger crowds may often be seen in very many places in Europe, but surely such a motley gathering as that which congregated on this occasion, can only be seen in the East. Imagine them all seated down and listening with wrapt attention to a yellow robed priest, holding forth from the platform filled with

Budhist priests, clergymen, and Singhalese clad in their national costume, and your readers can form some idea—a very faint one indeed—of the heterogeneous mass that revelled in a display of Singhalese eloquence seldom heard in this country.[1]

The coastal regions of the Buddhist kingdom of Sri Lanka had been conquered by the Portuguese in the early sixteenth century; in 1592, in order to escape the Portuguese, the royal capital was moved to Kandy in the highlands, the site of the most sacred relic on the island, a tooth of the Buddha. Roman Catholic missions were established, seeking to convert the Sinhalese in the lowlands. In 1638 the Portuguese were attacked by the Dutch, who eventually gained control of the entire island, apart from the kingdom of Kandy. They held the coastal regions until 1796, when they were displaced by the British. In the 1802 Treaty of Amiens between France and Britain, Napoleon (who controlled the Netherlands) formally ceded control of Sri Lanka to the British. By 1815, and after two bloody wars, the British controlled the entire island. Sri Lanka would remain the British crown colony of Ceylon until 1948. Under the British, a number of Protestant missions were established in the nineteenth century, seeking to convert the Buddhist populace to Christianity. They achieved a certain degree of success.

In 1862 a Buddhist monk named Guṇānanda founded the Society for the Propagation of Buddhism and established his own printing press, publishing pamphlets attacking Christianity. A number of Wesleyan clergymen responded to his charges, both from the pulpit and in print. And so in 1873, a public debate between Guṇānanda and a Christian representative, Rev. David da Silva (a Sinhalese convert), was arranged.

In their speeches (each party was allotted one hour in both the morning and the afternoon sessions each day), the adversaries sought to demonstrate the fallacies of the other's doctrines and scriptures. The Reverend da Silva spoke first, making extensive references to the Pāli *sutta*s and what the reporter from the *Ceylon Times* called "the abstruse metaphysics of Budha." His first target was the doctrine of no self, that the person is only an aggregation of various impermanent constituents. According to Buddhism, he said, human beings have no immortal soul and are "on a par with the frog, pig, or any other member of the brute creation."[2] Furthermore, if there is no soul there can be no punishment for sin and no reward for virtue in the next life, and thus no motivation to seek the good and shun evil. "What villain would not exult in the idea that he is not to suffer for what he does in this life!" Thus, "no religion ever held out greater inducements to the unrighteous than Buddhism did."[3]

The Buddhist monk Guṇānanda then rose to speak. He was described by the *Ceylon Times* reporter as "a well-made man of apparently forty five or fifty years, rather short, very intellectual looking, with eyes expressive of great distrust, and a smile which may either mean profound satisfaction or supreme contempt."[4] He declared that Rev. da Silva's recitations of passages from the Buddhist scriptures were filled with blunders in pronunciation; there was little reason to expect that the reverend had understood something as profound as the Buddha's teachings on the nature of the person. He then began to enumerate the contradictions that occur in the Bible. He noted that in Genesis, God regrets having created man and asked whether it was the omniscient creator or the fool who regrets his deeds. In Exodus God instructs the Hebrews to mark their doors with blood so that he will know which houses to pass over as he kills the firstborn of the Egyptians; "if he were omniscient, surely this was not necessary."[5]

In response, his Christian opponent alluded to the story of Prince Vessantara, the famed apostle of the perfection of giving, who, in one of the most poignant scenes in Buddhist literature, gave away his children and then his wife. In his next life, Prince Vessantara was reborn as Prince Siddhārtha, who became the Buddha. The clergyman asked the audience, "Were these meritorious acts? Was it meritorious to break the hearts of wives and children, and bring desolation and misery to a happy home? If it were, what actions will they enumerate under the head of demerits or sins?"[6] And so the debate continued over to the afternoon and the next day. At the conclusion of the event, Guṇānanda was declared the winner by the acclamation of the audience.

The debate, and the history of its representation, are fascinating, and deserve far more attention than can be provided here. But I would like to note one other exchange between the two parties, who are identified in the *Ceylon Times* article as "the Priest" (that is, the Buddhist monk) and "the Catechist" (that is, the Protestant clergyman). On the afternoon of the second day, Rev. da Silva described Mount Meru, the square mountain that, according to the Buddha's description, occupies the center of our universe. It is said to be 80,000 *yojana*s high, 80,000 *yojana*s wide, and 80,000 *yojana*s long. (A *yojana* was a unit for measuring distance in ancient India, derived from the distance that a pair of yoked oxen could travel in one day; it was apparently considered to be equal to approximately 16 miles in Sri Lanka in the nineteenth century.) Rev. da Silva asked, "How is it possible, that it [Mount Meru] could not be seen by the eyes of men?" According to one account, a Buddhist monk, presumably referring to the account in Genesis of the Tree of the Knowledge of Good and Evil in the Garden of Eden, shouted from the crowd. "Climb to the top of the tall tree described in your sūtras and you will definitely see it."[7] After the laughter died down, Rev. da Silva held up a globe.

> In this the shape of the earth, its dimensions, the great rivers and seas, and the positions of the countries, &c., are all represented. Now the circumference of the earth is 25,000 miles. This is admitted by all the civilized nations of the world. This fact is proved by every day's experience. Therefore a mountain with such dimensions could not exist on this earth. Wherever it existed it must be seen, as this globe which now stands on this little inkstand, must be seen by all who are on the four sides of it. So likewise if there were a mountain of that kind it could not but be seen by all the inhabitants of the four quarters. Besides, man can know to a certainty within a few weeks whether there be such a mountain or not. Men at no period ever saw such a mountain, nor have they known by science that there could be such a mountain. One who said that there was such a mountain cannot be supposed to have been a wise man, nor one who spoke the truth.[8]

The question of the location of Mount Meru specifically, and the Mount Meru cosmology more generally, is one that has vexed Buddhists, and their opponents, for centuries. Before hearing what "the Priest" said in response, let us briefly describe the standard Buddhist cosmology.

Classical Buddhist cosmology describes multiple universes that pass in and out of existence over four cosmic phases (each of which is twenty aeons[9] in length), called nothingness, creation, abiding, and destruction. After a period of nothingness, the physical universe comes into being during the period of creation, which begins when the faint wind of the past karma of beings begins to blow in the vacuity of space at the end of the previous period of nothingness. Beings come to inhabit the world during the period of abiding. During the period of destruction, the physical universe is incinerated by the heat of seven suns. This is followed by a period of nothingness as the fourfold cycle begins again.

The various Buddhist traditions of Asia generally subscribe to a map of the world presented, among other texts, in the *Abhidharmakośa* (Treasury of Knowledge) by the fourth-century Indian scholar Vasubandhu. After the period of cosmic vacuity mentioned above, this is how a world (*lokadhātu*) forms: At its foundation is a vast circle of wind, surmounted by a vast circle of water, surmounted by a vast circle of golden earth. In the center of that earth is a great mountain, called Meru or Sumeru. It is surrounded by seven mountain ranges of gold, each separated from the other by a sea. At the foot of the seventh range, there is a great ocean, contained at the distant perimeter of the world by a circle of iron mountains. In this vast ocean, four island continents are situated in the four cardinal directions, each flanked by two island subcontinents. The northern continent is square, the eastern continent is semicircular, the southern continent is triangular, and the western continent is round. Although humans inhabit all four continents, the "known world" is the southern continent, called Jambudvīpa, where the current average height is 4 cubits and the current life span is one hundred years; on the northern continent of Uttarakuru the average height is 32 cubits and the inhabitants live for one thousand years. The four faces of Mount Meru are flat, and each is composed of a different precious stone: gold in the north, silver in the east, lapis lazuli in the south, and crystal in the west. The substance determines the color of the sky for each of the four continents. The sky is blue in the southern continent of Jambudvīpa because the southern face of Mount Meru is made of lapis.

According to Buddhist doctrine, the beings who wander in saṃsāra, the realm of rebirth, are of six types: gods, demigods, humans, animals, ghosts, and hell beings, each of which has a place in this world system. Gods are of three types: those of the Realm of Desire (*kāmadhātu*), those of the Realm of Form (*rūpadhātu*), and those of the Formless Realm (*ārūpyadhātu*). There are six types of gods in the Realm of Desire, each with a different heaven. The gods of the Four Royal Lineages inhabit the upper reaches of the four slopes of Mount Meru. The gods of the Heaven of the Thirty-Three inhabit its summit. The other four types of gods in the Realm of Desire inhabit celestial realms at differing heights above Mount Meru. The Realm of Desire encompasses Mount Meru as well as the four heavens above it, the great ocean, the four continents, and the hells beneath (described below).

Above the Realm of Desire is the Realm of Form, a system of heavens reserved for those who have achieved deep states of concentration in their previous life. It is divided according to the level of concentration reached previously, and thus enjoyed in these heavens. There are four major divisions, called (in ascending order) the First Concentration, the Second Concentration, the Third Concentration, and the Fourth Concentration. Each has various divisions, three in each of the first three and eight in the fourth. The deities who inhabit these heavens experience only three of the five sense objects: those of visible form, sound, and touch.

Beyond the Realm of Form is the Formless Realm, not a place because it is not physical, but still a place of rebirth for those who have attained the deepest levels of concentration in their previous life. Here the beings have no bodies, only minds, which contemplate four objects that provide the names of the realms: Infinite Space, Infinite Consciousness, Nothingness, and Neither Existence nor Non-Existence.

Of the other five types of beings in saṃsāra, the demigods (*asura*) inhabit the lower slopes of Mount Meru. Humans are found on the four islands surrounding it (although whether someone 32 cubits tall who lives for one thousand years is "human" would seem to be a question). Animals occupy the four continents, the skies above them, and the oceans that surround them. Ghosts are said to inhabit a realm beneath the ground as well as regions of the southern continent. Buddhist texts describe an elaborate system of eight hot hells and eight cold hells, as well as neighboring hells, all located at various depths beneath the surface of the earth; it is noteworthy that the hells are located not directly below Mount Meru, but beneath the southern continent, our continent, of Jambudvīpa.[10]

According to a widely known creation account, the first humans in the present period of abiding had a life span of eighty thousand years. They descended from the Formless Realm and the upper levels of the Realm of Form (which are not destroyed by the seven suns during the phase of destruction). Free from the marks of gender, they were able to fly and were illuminated by their own light; there was no need for a sun or moon. They also did not require food. At that time, the surface of the earth was covered by a white frothy substance. One of the beings descended to earth and dipped a fingertip into the substance and then touched the finger to its tongue. The taste was sweet. Soon all the beings began eating the white substance, which would naturally replenish itself. But the introduction of this food into their bodies soon caused them to lose their natural luster, and the sun and moon appeared to illumine the sky. The added weight of their bodies soon made it impossible for them to fly. The white substance evolved into a naturally growing huskless rice that would be ready to harvest again the day after it was picked. But as the beings ate more and more of the rice, it became necessary for them to somehow eliminate the waste that was accumulating in their bodies, and the anus and genitals developed. One couple soon discovered an additional use for the genitals and engaged in sexual intercourse for the first time. The others were scandalized, pelting them with mud. Soon, to hide their shameful activities, people began to build houses. Growing too lazy to pick the rice each day, they began to take more than they needed and hoard it in their houses. As a result, the rice developed husks and required more and more time to grow. Soon people began to steal from each other, requiring the election of a king who would enforce a system of laws. And this is how human society came into existence in this world.

Thus, setting aside for the moment the heavens above and the hells below—crucial though they are to Buddhist doctrine and practice—it can be said in summary that the human realm that Buddhist texts describe is a flat earth, or perhaps more accurately a flat ocean, its waters contained by a ring of iron mountains. In that ocean is a great central mountain, surrounded in the four cardinal directions by island continents.

This cosmography, with Mount Meru at its center, would provide the site for the first encounters between Buddhism and Science, encounters not of compatibility but of conflict. In 1552, more than three centuries before the Pānadurē debate, Francis Xavier described his mission to Japan:

> And for the greater manifestation of God's mercy, the Japanese are more subject to reason than any other pagan race that I have ever seen. They are so curious and importunate in their questioning and so eager to know that they never ceased asking us questions and telling others the answers which they had received from us. They did not know that the world was

round, nor did they know the course of the sun. They asked about these and other things, for example, about comets, lightnings, rain and snow, and similar phenomena. They were very content and satisfied with our replies and explanations; and they deemed us to be learned men, something that was of some help in gaining credit for our words.[11]

It is important to note that the spherical earth that the renowned Jesuit saint describes remains the center of the universe; he refers here to "the course of the sun." The church had yet to accept a heliocentric universe; Copernicus had published his *De revolutionibus orbium coelestium* just ten years earlier. That the earth was round would not be understood by the leading Japanese astronomers until the late eighteenth century, thus challenging the traditional Buddhist view of a flat world. For at least one Buddhist intellectual of the period, however, the size and location of Mount Meru were of little consequence.

The brilliant Tokugawa scholar Tominaga Nakamoto (1715–1746) was among the first scholars in the world, whether Asian or European, to offer what would be considered today a "historical" analysis of the development of Buddhism. It is an article of faith in the Mahāyāna traditions of India, China, Korea, Japan, and Tibet to regard the vast corpus of Mahāyāna sūtras (including such famous works as the *Lotus Sūtra*, the *Diamond Sūtra*, and the *Heart Sūtra*) as the word of the Buddha. Indeed, in some versions of the bodhisattva's vows, it is a transgression not to do so. Beginning in the nineteenth century, however, European (and later, Japanese) scholarship demonstrated that these sūtras first began to appear some four centuries after the death of the Buddha. Long before this European "discovery," however, Tominaga argued that only a small portion of the words ascribed to the Buddha were indeed his, and that it was not possible to harmonize the apparently contradictory teachings found in the vast canon under the hermeneutical scheme of a single school that claimed to represent the Buddha's own view. Instead, he saw the proliferation of Buddhist sūtras as evidence of contention, with competing factions composing their own sūtras in the centuries after the Buddha's death, and each faction ascribing its sūtra to the Buddha himself in an effort to claim his authority.

Among the numerous contradictions that Tominaga discovered in the Buddhist scriptures was a vast range of opinion on the size of Mount Meru. For him, this was simply further evidence of his view of the history of Buddhism. He assumed that at least one of these descriptions of Mount Meru had been provided by the Buddha. However, unlike several important Japanese Buddhist authors of the subsequent century, Tominaga was unconcerned that the cosmography described by the Buddha might be wrong. In his *Emerging from Meditation* (*Shutsujōkōgo*), composed between 1735 and 1737, he writes:

> The teachings about Mount Sumeru were all handed down by brahmans. Though Śākyamuni used them to expound the Way, they are to be regarded as cosmic theory. Later scholars however have made much of this teaching while criticising others, and lost sight of the Buddha's intention. This is because the Buddha's intention is not to be found in such matters. He was urgently seeking people's salvation and had no time for such petty matters. What he did is what is known as skilful means. . . . Those who discuss the rights and wrongs of such things are all small-minded people. Recently there have been people who have indulged in taking this up again when putting together theories about the universe. It is extremely squalid, indeed ridiculous. . . . Teachings about the cosmos are in actuality quite vague and do no more than tell us of the inner workings of the mind. There is no way of knowing whether they are right or wrong. Hence I say that the cosmos arises on the pattern of people's minds.[12]

His argument, which we will encounter elsewhere, is that the Buddha simply made use of the prevailing cosmography of his time, one that had been created by the brahman priests of ancient India, using it as a convenient setting, no more than a backdrop, for his exposition of the path to liberation from suffering. The details of this cosmography were not of the Buddha's making, and, in fact, he had no particular interest in them. Hence, whether this cosmography is right or wrong is irrelevant because the Buddha had no investment whatsoever in its truth; it is the height of folly to claim that the Buddha was mistaken because he described a world that does not exist.

Other Japanese Buddhist thinkers, however, sought to defend the Mount Meru cosmography against its critics, both European and Japanese. Perhaps the most interesting of these was Entsū Fumon (1755–1834), a Buddhist monk of uncertain affiliation (he is variously described as belonging to the Nichiren, Tendai, and Jōdo sects) and author of the five-volume work *Astronomy of the Buddhist Country* (*Bukkoku rekishōhen*), published in 1807. Heliocentric theory had been introduced into Japan in the late eighteenth century, through the translation of Dutch astronomical texts and the composition of Japanese works based on these translations. This occurred as part of the Japanese assimilation of European arts and sciences, known as "Dutch Learning" (*rangaku*). Entsū studied this astronomy and set out to use its methods to demonstrate the accuracy of the traditional Buddhist cosmography, which he referred to as the *shumisensetsu*, the "Mount Sumeru theory." He called the object of his research *bonreki*, "Indian astronomy" (although he was unaware of the work of the great sixth-century mathematician and astronomer Āryabhaṭa), and *butsureki*, "Buddhist astronomy." His claim was that the traditional Buddhist description of the universe was not contradicted by European astronomy. He argued instead that astronomical and meteorological events are better explained under the Buddhist model, and that the heliocentric universe with its spherical earth described by European astronomy is in fact a misinterpretation of scientific data; the correct interpretation of that data demonstrates the existence of a universe as it is described in Buddhist texts. He attracted a large following, but one that dissipated not long after his death, leaving him an almost forgotten figure.

Although Entsū's work is filled with detailed calculations, his ultimate appeal is to the omniscience of the Buddha; the universe is described accurately in Buddhist texts because those texts derive from the teachings of the Buddha, who is endowed with a divine eye (*tengen*) that is not possessed by other beings. Thus, the eye of the Buddha perceives things that remain invisible to the human eye. The Buddha also had full knowledge of the past, present, and future and set forth the Mount Meru cosmography to counter the belief in a spherical earth, a belief he knew would arise in the future.[13]

Globes showing the continents of the world were well known in Japan at this time. Entsū therefore needed to account for these continents within the Buddhist system. He took as his primary source the description of the world in the *Abhidharmakośa* (summarized above). However, evincing a certain empiricism absent in other Japanese Buddhist apologists of the day, he did not consider the heavens above Mount Meru or the hells below. Instead, he limited his description to the "human" world, that is, the world that can be seen both by the divine eye of the Buddha and the fleshly eye of humans. It was Entsū's task, therefore, to correlate the continents in the four cardinal directions around Mount Meru (each of which is flanked by two subcontinents) with the landmasses that appeared on European maps and globes. Thus, Asia, Europe, and Africa together constituted the southern island of Jambudvīpa, which he argued had once been a single landmass; its triangular shape, narrow at the south and wide at the north (as described in Buddhist texts), remains visible. America and Antarctica are the two subcontinents flanking Jambudvīpa.

Much of Entsū's book is devoted to astronomy. According to the Mount Meru system, the sun and moon are supported by a ring of wind that floats in the sky halfway between the surface of the great ocean and the summit of Mount Meru. The sun orbits Mount Meru each day in a clockwise direction, such that when it is daytime in the southern continent of Jambudvīpa, it is sunrise in the western continent, night in the northern continent, and sunset in the eastern continent. The difference in the length of the day over the course of the year is explained by the fact that the sun moves north and south over the southern continent. However, this explanation would not account for the fact, known in Japan, that there are parts of the world that spend half the year in daylight and half the year in darkness. Entsū thus goes to considerable lengths to account for the existence of what he calls the "dark country" (*yakoku*). He explains that during the winter season on the southern continent, the orbit of the sun in its circuit around Mount Meru shifts to the south. The orbit of the sun consequently shifts closer to the northern face of Mount Meru as it passes across the northern continent. This reduces the amount of daylight, and hence the length of the day, during winter in the southern continent. There is also a mountain range that runs horizontally across the southern continent of Jambudvīpa, which Entsū calls the Kunlun (Kanron), referring presumably to the range that runs across northern Tibet (although on his map he placed it between Africa and Asia). When the sun passes to the south of this range during the wintertime, these mountains block the sunlight from reaching the region to their north. Hence, during the wintertime, the area to the north of the Kunlun range has its sunlight blocked by Mount Meru in the north and by the Kunlun mountains to the south, making it the "dark country."[14]

Entsū provided detailed maps with precise measurements of the continents and subcontinents of the Buddhist world. As a Mahāyāna Buddhist, he held that the universe is ultimately formless, and thus beyond measurement. However, because the divine eye of the Buddha can perceive this formless universe, and because the Buddha then described in the sūtras what he had seen, it is possible to produce maps and models of the universe—maps and models that were confirmed, in Entsū's view, by European astronomy and mathematics. In fact, he does not argue that European science confirms the vision of the Buddha, but that the Buddha's vision confirms European science.[15] He went so far as to make a mechanical scale model of the Mount Meru world, one that reflected the size and location of the mountains, continents, islands, and oceans, as well as the size and motion of heavenly bodies, as they are described in Buddhist texts. A version of this model was prominently displayed at the first Japanese exposition, held in Tokyo in 1877.

Entsū's writings were widely read in the nineteenth century among Buddhists engaged in anti-Christian polemics; several of the major Buddhist seminaries in Japan, especially those of the Pure Land sects, established departments of astronomy (*rekigaku*), where the flat-world Mount Meru cosmography was studied. However, the Buddhists of Japan did not univocally support the traditional cosmology. Some clung fiercely to Mount Meru, claiming that the cosmography was a literal and original teaching of the Buddha (not derived from the brahmans), and that to abandon the existence of Mount Meru is to abandon the law of karma.[16] Others, however, came to adopt the view that Tominaga had espoused more than a century earlier.

One such figure was the prominent Jōdo Shinshū priest Shimaji Mokurai (1838–1911). He argued that those who condemn the round-world cosmography of the Europeans and defend the flat-world cosmography of the Buddhist texts in fact do a disservice to the dharma; the Mount Meru cosmography is not the foundation of Buddhism. In fact, Mount Meru is simply an element of Indian myth, like Mount Olympus in Greek myth; no one in the modern world believes that they are real. Appealing to Buddhist doctrine, he notes that the Mount Meru cosmography provides the setting for the doctrine of transmigration of sentient beings in saṃsāra. However, the Mahāyāna sūtras consistently describe saṃsāra as an illusion. To claim that Mount Meru exists is to attempt to turn that illusion into a reality. Referring to those who sought to save Mount Meru, he wrote, "Of course, I am deeply impressed by their sincerity to preserve the dharma, but the foundation of Buddhism is unfortunately not on this issue."[17]

The Jōdo Shinshū educator and philosopher Inoue Enryō (1858–1919) argued that the Mount Meru cosmography was a Hīnayāna teaching, and thus was ancillary to Buddhism; whether it is true or not is immaterial, although it remains of historical interest.[18] The Sōtō Zen priest Kimura Taiken (1881–1930), reflecting the views of European Buddhology, took the opposite view. He sought to distinguish the views of the Buddha, which he called "original Buddhism" (*genshi bukkyō*), from the popular accretions that had polluted the tradition in the centuries after the Buddha's death. Original Buddhism was scientific; unscientific elements entered with the rise of the Mahāyāna. It was necessary to employ original Buddhism to purify the Mahāyāna in order to create a "new Mahāyāna movement" (*shin daijō undō*). Among the unscientific elements that required purification was the Mount Meru cosmography; according to Kimura, it should not be regarded as a teaching of the Buddha. Rather, the Buddha found it expedient to make use of ancient Indian beliefs about the world as a context for his own teachings; his ultimate intention was not to describe this world. Thus, the Mount Meru cosmography should not be ascribed to the Buddha; it is in fact a Brahmanical view. The teaching of the Buddha should not therefore be condemned because it includes the description of a world that has proved not to exist. If the Buddha had lived in the modern period, he would have described the world based on current scientific knowledge.[19]

The view of Meru as metaphor eventually came to be widely accepted by the Buddhist thinkers of Japan. Entsū's elaborate cosmography enjoyed popularity at a time when it could be seen as Buddhist science, one that could compete with the newly introduced European science. But the European scientific worldview soon came to be adopted by the Meiji government; the Ministry of Doctrine (Kyōbushō), which only lasted from 1872 to 1877, prohibited the teaching of the Mount Meru cosmography. Under such circumstances, Entsū's detailed argument that the Buddha had indeed meant what he said would soon fall into oblivion.

The existence of Mount Meru, or at least its description, had been known to Europeans since at least the seventeenth century. One of the more detailed accounts was provided by the French Jesuit Guy Tachard (1651–1712) in a work entitled (in its 1688 English translation) *A Relation of the Voyage to Siam: Performed by Six Jesuits, Sent by the French King, to the Indies and China, in the year, 1685: With their Astrological Observations, and their Remarks of Natural Philosophy, Geography, Hydrography, and History Published in the Original, by the Express Orders of His Most Christian Majesty; and Now made English, and Illustrated with Sculptures.* Father Tachard presumably derived his description from conversations with Thai monks.

> The Earth, in their Philosophy, is not round, it is only a flat Surface; they divide it into four square parts, which they call *Thavip*. The Waters, by which those four parts are separated, not being navigable, because of their extream subtility, hinder the commerce that they might have with one another. The whole Earth is encompassed with an extreamly strong and very high Wall. On this Wall all the Secrets of Nature are engraven in great Characters, and there it is that these wonderful Hermits whom I mentioned, learn all the admirable things they know; for they easily conveigh themselves thither with that surprizing agility they are endowed with. As to the men of the other three Parts of the World, they have a Countenance much different from ours; for the Inhabitants of the first have a square Face, of the second a round, and of the third a triangular....
>
> In the middle of the four Parts of the World there is an exceeding high Mountain, called in *Siamese Ppukhan Pprasamen*. It rests upon three precious Stones, very little ones, its true, but strong and solid enough to support it. Round this Mountain the Sun and Moon continually turn, and by the revolution of those two Luminaries, Day and Night are made. This great Mountain is environed by three Rows of lesser Hills, of which, there is one all of Gold. The great Mountain is inaccessible, because the Water that surrounds it is not navigable. As for the Mountain of Gold, a fearful Gulf renders the approach to it most difficult. It is true, a rich man heretofore got to it, but it was with extream danger of being lost in that Abyss, whither all the Waters come and muster, and from whence afterward they gush out to make the Sea and Rivers.
>
> The whole Mass of Earth hath underneath it a vast extent of Waters, which support it as the Sea bears up a Ship. These inferior Waters have a communication with those that are upon the Earth, by means of the Gulf I have been speaking of. An impetuous Wind holds the Waters under the Earth suspended, and this Wind, which exists of it self, and has no cause, blowing from all eternity with incredible violence, drives them continually back, and hinders them falling. When the time is come that the God of the *Siamese* hath foretold, that he shall cease to reign, then the Fire of Heaven falling upon the Earth, shall reduce into Ashes everything that

> comes in its way, and the Earth being so purified, shall be restored again to its former state.[20]

This apparently sympathetic rendering of the Buddhist world, marked by the European voyager's interest in both gold and navigable waters, does not, however, imply a particular sympathy for Buddhism (although the term did not exist in a European language in the seventeenth century). A few pages later, Father Tachard has this to say about the Buddha, whom he refers to as *Sommonokhodom*, his rendering of the Thai pronunciation of the Buddha's epithet, Śramaṇa Gautama, the ascetic Gautama:

> I thought fit to premise all these things before I came to speak of *Sommonokhodom* (so the *Siamese* call the God whom at present they adore) because they are necessary to the understanding of this History. That History, after all, is a monstrous mixture of Christianity and the most ridiculous Fables. It is at first supposed that *Sommonokhodom* was born God by his own virtue; and that immediately after his Birth, without the help of any Masters, to instruct Him, he acquited by a meer glance of his Mind, a perfect knowledge of all things relating to Heaven, the Earth, Paradice, Hell, and the most impenetrable Secrets of Nature; that at the same time he remembred all that ever he had done in the different Lives he had led; and that after he had taught the People those great Matters, he left them written in Books, that Posterity might be the better for them.[21]

The location and existence of Meru would be regarded as a ridiculous fable by Christian missionaries over the next two centuries. It was a particular target in Sri Lanka in the early nineteenth century. Here the stakes were high. The missionaries felt that if they could debunk Buddhist myths with modern scientific knowledge, they could convert disillusioned Buddhists to the Christian faith. In 1816 two recent Sinhalese converts from Buddhism to Christianity were taken to England for their studies. It is reported that their faith in Buddhism (and its cosmography) was shaken not by the gospel but by the disproportionate length of day and night in England, so different from Ceylon, located near the equator.[22]

The traditional Buddhist cosmology was thus studied by Wesleyan missionaries in the British colony of Ceylon in the early nineteenth century; they then used it, repeatedly, to demonstrate the fallacies and inadequacies of Buddhism. Among these missionaries to the island, Daniel George Gogerly (1792–1862) and Robert Spence Hardy (1803–1868), both of the Wesleyan Methodist Missionary Society, took the lead in describing and debunking Mount Meru in some detail.[23]

Their argument, in summary, was that the European (that is, Christian) description of the universe was right, and the Buddhist description was wrong. This championing of the heliocentric universe by Christian clerics for the conversion of idolaters (as Buddhists were regarded) perhaps seems ironic in light of the notoriously strained relations between certain astronomers and the church in previous centuries. In Sri Lanka, however, it was regarded as a powerful weapon, because it could be used to demonstrate empirically the errors of the Buddha.

That Mount Meru does not exist could be demonstrated from any number of perspectives. Hardy provides such an argument in his *The Legends and Theories of the Buddhists Compared with History and Modern Science*, published in English in 1863 and in Sinhalese in 1865:

> An objection is sometimes raised by the Buddhists, that as there are some parts of the world not yet visited by Europeans, these parts, if visited, might prove that the Buddhist is right, and the European wrong. But this cannot be. There are probably some parts of the province of Bintenne [in Sri Lanka] not yet explored by the white man; but he has been on every side of it, and knows that it can only be of a certain size; and it is the same with other unvisited lands; he has been all around them, and can tell exactly their extent. Except the interior of Africa and Australia, and the north and south poles, nearly every part of the earth has been seen by the eye of the traveller or navigator; and if we could sift the evidence they would give, under an examination on these subjects by a board of scientific men, it would all tend to prove that Buddha was ignorant of the true figure of the earth, and that all he says about it is unscientific and false.[24]

The Buddhists did not remain silent. In 1839 a Buddhist author had written a rejoinder to a polemical tract composed by Hardy (no longer extant). In it he raises a series of familiar objections against a round and rotating earth: why don't cups and saucers fall off tables and water fly from lakes and wells?[25] Gogerly responded to these and other objections in his "Kristiyāni Prajñapti" ("The Christian Teaching"), first published in

Sinhalese in 1848. There he explains that the world is round and not flat because a ship leaving Ceylon and sailing west does not eventually reach the ring of mountains that contains the vast ocean, as the Buddhists would hold, but in fact eventually reaches Ceylon.[26] The prominent Buddhist monk Bentara Atthadassī (d. 1862) wrote an extensive rejoinder to Gogerly, pointedly entitled "Bauddha Prajñapti" ("The Buddhist Teaching"). He was contemptuous of the British claim that Mount Meru did not exist because no one had ever seen it; the continent of Jambudvīpa is so vast that supernormal powers are required to reach Mount Meru.[27]

The Christian missionaries were unmoved by such claims. Indeed, Hardy focused on the Buddha's geographical error as a fatal flaw.

> There can be no doubt that Buddha taught the existence of Maha Méru.... An attempt may be made to set aside the consequences of this exposure of Buddha's ignorance, by saying, that this is a kind of mistake that does not invalidate his doctrines; Buddhism may still be true as a religious system. But this is a fallacy that I am most anxious to set aside. If Buddha said that which is false, under the supposition that it is true, he betrays ignorance, imperfect knowledge, and misapprehension. He cannot, therefore, be a safe teacher; there may be some things about his religion that are true, as there are about every religion; but it is not a revelation; its author was a mere man, with limited and imperfect knowledge; and to receive it as the pure unmixed truth, is a mischievous and fatal mistake.[28]

This brings us back to 1873 and the debate between Rev. da Silva and the Buddhist monk Guṇānanda. As we recall, da Silva had held up a globe and asked why Mount Meru did not appear on it, coming to the same conclusion that Hardy had reached a decade before. Da Silva said, "Men at no period ever saw such a mountain, nor have they known by science that there could be such a mountain. One who said that there was such a mountain cannot be supposed to have been a wise man, nor one who spoke the truth."[29]

Guṇānanda did not offer the predictable answers: that Mount Meru was located in an unmapped territory or that one required supernormal sight to see it. Instead, he attacked the Christian's science.

> The Revd. gentleman no doubt alluded to Sir Isaac Newton's theory when he made that remark, according to which day and night were caused by

the earth revolving round its axis, and not by the sun being hidden behind *Mahāmeru*. The little globe too which the Revd. gentleman produced was also one made on Newton's principle; but even amongst Englishmen there were serious doubts and differences of opinion as to whether Newton's theory was correct or not. Among others one Mr. Morrison, a learned gentleman, has published a book completely refuting all Newton's arguments, and he would be happy to allow the Christian party a sight of this book which was in his possession. . . . How silly was it then to attempt to demolish the great Buddha's sayings by quoting as an authority, an immature system of Astronomy, the correctness of which is not accepted even by those who propounded it.[30]

Guṇānanda refers here to the work of Richard James Morrison (1795–1874), a lieutenant in the Royal Navy who had gone on to become the most famous astrologer of Victorian England, publishing predictions of weather and world events under the pseudonym "Zadkiel" or "Zadkiel, Tao-Sze."[31] His works include *The Solar System As It Is, And Not As It Is Represented: Wherein is Shewn, for the First Time, the True, Proper Motion of the Sun through Space, at the Rate of 100,000 Miles Per Hour. Also, That the Earth and Planets, and their Satellites Move with the Sun, in Cycloidal Curves; and That the Doctrine of Elliptical Orbits is False* (1857). Guṇānanda seems to be referring to another of Morrison's books: *The New Principia; or, True System of Astronomy. In which the Earth is Proved to be the Stationary Centre of the Solar System, and the Sun is Shewn to be Only 365,006.5 Miles from the Earth, and the Moon Only 32,828.5 Distant; While the Sun Travels Yearly in an Ellipse around the Earth, the Other Planets Moving about the Sun in Ellipses Also* (1868). It is noteworthy that this argument against Newton and in favor of a geocentric universe was sufficiently successful to warrant a second edition in 1872. But Guṇānanda does not enter into the specifics of scientific evidence here. For him, it is simply enough to note that even the English cannot agree on whether the earth rotates on its axis in orbit around the sun. Therefore, why should the Buddhists of Sri Lanka yield to the view of one Englishman (Newton) over another (Morrison)? Guṇānanda seems unencumbered by the question of the relative importance of the two figures in the history of science.

Not satisfied with calling the authority of Newton into question, Gunāṇanda turned next to empirical evidence. According to Buddhist cosmology, we humans occupy an island continent called Jambudvīpa, located to the south of Mount Meru. If "the grandest and most stupendous rock on the face of the earth" is not situated to the north, how does the Reverend gentleman explain the invariable direction of the mariner's compass?[32] We should recall that Guṇānanda won the debate.

The location of Mount Meru had been a question for Buddhists in China for centuries, both before and after the arrival of Christian missionaries. And by the late nineteenth century, educated Chinese accepted the European view of a round earth orbiting the sun. In 1927 the monk Taixu (discussed in the introduction) published his *Essay on the True Nature of Reality in Accordance with the Teachings* (*Zhen xianshi lun zong yi lun*), in which he sought to harmonize the Buddhist and European cosmologies. He argues that the Mount Meru cosmology is in fact a metaphor for the solar system. Mapping the elements of the Buddhist system onto the stars and planets, he explains that the southern continent of Jambudvīpa is the planet Earth and that the other three islands in the ocean surrounding Mount Meru are Venus (Godānīya), Mercury (Pūrvavideha), and Mars (Uttarakuru). Each of these is inhabited by humans. Saturn, Jupiter, Neptune, and Uranus are inhabited by various gods. He is less specific about the precise location of Mount Meru. He does say, however, that the sun is located directly above Mount Meru. In the standard Buddhist cosmology, as discussed above, the summit of Mount Meru is the site of the Heaven of the Thirty-Three, presided over by Śakra (Indra). Taixu explains that this heaven is located on the sun.[33]

After having considered Sri Lanka, Japan, and China, we turn now to the status of Mount Meru in yet another Buddhist land, Tibet. As the current Dalai Lama himself has noted, in the case of Tibetan Buddhism, "for various historical, social, and political reasons, the full encounter with a scientific worldview is still a novel process."[34] Indeed, the Mount Meru question was not raised in Tibet until the twentieth century. The Tibetan scholar Gendun Chopel (Dge 'dun chos 'phel, 1903–1951) spent the years from 1934 to 1946 traveling in South Asia. During this period, he contributed poems and essays to *Melong* (its English name was the *Tibet Mirror*), the only Tibetan-language newspaper, published in Kalimpong in northern India by the Tibetan Christian Babu Tharchin (1890–1976).

In 1938 Hitler annexed Austria; Otto Hahn produced the first nuclear fission of uranium; Howard Hughes, flying a twin-engine Lockheed, set a new record for the circumnavigation of the globe; color television was first demonstrated; the first photocopied image was produced; the ballpoint pen was patented; the first *Superman* episode appeared in Action Comics; Disney's *Snow White and the Seven Dwarfs* premiered; Benny Goodman's orchestra performed "Sing, Sing, Sing" at Carnegie Hall—and Gendun Chopel was attempting to prove to his countrymen that the world is not flat.

The following essay appeared in the June 28, 1938, issue of the *Tibet Mirror*, under the byline "Honest Dharma."

The World Is Round or Spherical

In the past, in the lands of the continent of Europe, it was said only that this world is flat, just as it appears to the non-analytical mind; there was not a single person who said that it was round. All the ancient religions in the various lands said only that the world is flat; there was not one that said that it was round. Thus, when some intelligent people first said that it is round, the only method to keep the word from spreading was to order that they be burned alive. However, in the end, unable to withstand the light of true knowledge, everyone came to believe that it is round. Today, not only has the fact that it is round been determined, but also the size all of the islands in the world just 4 or 5 *yojana*s long have been measured down to spans and cubits. Therefore, in the great lands there is not a single scholar who has even a doubt.

Among all of the Buddhists in Singhala [Sri Lanka], Burma, Siam, Japan, China, and so forth, there is not one who says that it is not true that it is round. Yet we in Tibet still hold stubbornly to the position that it is not. Some say mindless foolish things, like the foreigners' sending of ships into the ocean is a deception. I have also seen some intelligent persons who understand [that it is round] but, fearing slander by others, remain unable to say so. When even the most obstinate European scholars, who do not believe in anything without seeing the reason directly, were not able to maintain the position that it is not round and accepted it completely, then it goes without saying that this stubbornness of ours will come to an end.

[Saying that the world is not round] because the Buddha stated that it is flat is not accepted as authoritative in other [non-Buddhist] schools and thus does not do a pinprick of damage [to their assertion that it is round]. Even with regard to the scriptures of our own [Buddhist] school, which does accept [the Buddha's statement] as authoritative, because the majority of the sūtras were set forth by the Buddha in accordance with the thoughts of sentient beings, even in this case, we do not know what is provisional and what is definitive. If he set forth even matters of great importance, such as emptiness and the stages of the path to liberation, in various types of provisional meaning in accordance with the thoughts of sentient beings, what need is there to discuss these presentations of environments and their inhabitants? During the lifetime of the Buddha, when it occasionally happened that the way that the monks ate their food did not accord with [the customs] of the time and place, causing slight concern among the laity, he would make a rule that it was unsuitable. At that time, throughout all the world, the words "[the world] is flat" were as famous as the wind. Thus, even if the Buddha had said, "It is round," whose ear would it have entered? Even if he had said so emphatically, it would have had no purpose, even if he had demonstrated it with his miraculous powers. Nowadays, at a time when [the fact that the world is round] has become evident to billions of beings, there are still those of us who say, "This is your deception." In the same way, I am certain that they would not have believed it, saying, "This is the magic trick of Gautama." If all of us would believe in this world that we see with our eyes rather than that world that we see through letters, it would be good.[35]

Prior to departing for India in 1934, Gendun Chopel had completed the monastic curriculum at Drepung monastery in Lhasa. There he had studied the *Abhidharmakośa* and the traditional Mount Meru cosmology still accepted in Tibet in 1938, the time of his article. His argument against the traditional "flat-earth" view is twofold.

The first part of the argument is one based on consensus. It is clear that he regards the Europeans as generally superior to Asians in matters concerning the empirical description of the external world. His readers should take some comfort, therefore, in learning that in the ancient past, all peoples, including the Europeans, considered the world to be flat, as it indeed appears to be. In addition, all of the religions of the world asserted that the world is flat. This belief, furthermore, was held so tenaciously that when it was first suggested that the world is not flat, those who made such a suggestion were executed. The light of truth, however, eventually dispelled the darkness of ignorance; and in the subsequent centuries the proof that the world is round has been upheld, and all bodies of land and water that constitute the world have been accurately measured. It is noteworthy here that Gendun Chopel employs almost the same language as Hardy in 1863: the white man has been everywhere and seen (and measured) everything, and the flat world surrounding Mount Meru has not been found. Unlike Guṇānanda, who sought to discredit both Christian scriptures and Christian science, Gendun Chopel here is prepared to embrace the conclusion of the scholars of "the great lands," by which he would seem to mean the nations of Europe, who are unequivocal in their belief that the world is round. The task, as we shall see below, is not to do battle with European science but to harmonize it with, and, in a sense, absorb it into, Buddhism.

His implication, of course, is that the Tibetans are part of the dwindling flat-world camp. One could object that the non-Buddhist nations may hold such a view, but the Buddhist nations, who traditionally have shared the Mount Meru cosmology described above, do not. But Gendun Chopel deflates this objection by reporting that the Buddhists of the rest of Asia, those of Sri Lanka, Burma, Siam, China, and Japan, all hold that the world is round. The Tibetans thus form a tiny and doomed minority. He concedes that there are in fact some Tibetans who know that the world is round but fear the criticism they would suffer by saying so.[36] As he wrote in one of his poems,

> Everything old is hailed as the way of the gods.
> Everything new is considered the conjuring of demons.
> Most wonders are considered simply bad omens.
> This is our tradition to the present day,
> The tradition of the Buddhist kingdom Tibet.[37]

But the obstinacy of the Tibetans is nothing compared to that of the Europeans, such that it is only a matter of time before even the Tibetans come to accept the truth that the world is round. Gendun Chopel suggests that their persistence in holding to the old and discredited position should be regarded as a source of national embarrassment.

Turning next to the question of the Buddha, he concedes that someone might argue that the world is flat because the Buddha said so. But Gendun Chopel observes that citing a statement of the Buddha as proof, a tactic that might be employed in the debating courtyard of a Tibetan monastery, would carry little weight in the wider world, where the Buddha is not regarded as an infallible authority on all matters; one cannot prove a point in the wider world with the simple reason, "because it was stated by the Buddha." Gendun Chopel is, however, a Buddhist, and so in the second part of the essay, he must take up the more difficult question: The Buddha taught that the world is flat, when in fact it is round. How could the omniscient Buddha have been wrong?

Here Gendun Chopel, like Tominaga in Japan two centuries earlier, alludes to the famous doctrine of *upāyakauśalya*, often translated as "skillful methods" or "expedient devices," according to which the Buddha always taught what was most appropriate and useful for a given audience, whether or not it was "true." Thus it is said, for example, that the Buddha taught that there is a permanent self (*ātman*) to those beings incapable of understanding that there is no self—despite the Buddha's own knowledge that, in reality, there is no self. This flexibility of the Buddha in adapting his teachings to a particular circumstance, combined with the vast number of texts attributed to the Buddha by the Mahāyāna, resulted in a wealth of contradictions among the statements of the Buddha, the omniscient teacher who must be free of contradiction. All schools of Buddhism, therefore, employed devices (some of which claimed to be recommended by the Buddha himself) to distinguish what the Buddha had said in accordance with the exigencies of the moment from what the Buddha in fact knew ultimately to be true. Among a number of categories deployed to draw this distinction,[38] Gendun Chopel mentions one of the most famous, that of the provisional (*neyārtha*) and the definitive (*nītārtha*). Although the parameters of these categories are widely interpreted, as Gendun Chopel uses the terms here, the provisional teachings are those statements by the Buddha that he made for the benefit of his audience but to which he did not himself subscribe, and the definitive teachings are those statements that represent the Buddha's own view on a given point of doctrine. Gendun Chopel notes that the majority of the Buddha's statements fall into the former category and that it is impossible for us to know with certainty which statements are provisional and which are definitive. He is well aware that there are in fact detailed instructions on how to distinguish the definitive from the provisional, but he also is aware that the various schools of Buddhist philosophy differ both on what those instructions are and on what constitutes the Buddha's own view. Gendun Chopel thus professes a certain agnosticism as to what is provisional and what is definitive.

He does note, however, that the Buddha spoke provisionally even on some of the most consequential questions in Buddhist thought, such as the nature of emptiness (*śūnyatā*) and the stages of the path to liberation from rebirth. As a proponent of the Madhyamaka school, Gendun Chopel would hold that the Yogācāra teaching that there is no external world is a provisional teaching, a lesser emptiness, which the Buddha taught to those who were temporarily incapable of understanding the true emptiness, the utter absence of any intrinsic nature. As a proponent of the Mahāyāna, Gendun Chopel would hold that the nirvāṇa of the arhat (one who has achieved liberation from rebirth), the goal of the Hīnayāna, is a provisional teaching, that in fact all beings will one day undertake and complete the bodhisattva's path to buddhahood. If the Buddha resorted to the use of provisional teachings on these doctrines, so central to Buddhist thought and practice, why would he not resort to the provisional on such a relatively minor point as the shape of the world? That is, nothing would prevent the Buddha from saying that the world is flat when he knew that it is round.

He notes correctly that the Buddhist monastic code, the vinaya, is filled with accounts of the Buddha making minor modifications in the prescribed etiquette of monks and nuns in deference to the mores of the laity. The laity of ancient India believed without question that the world is flat. If the Buddha had taught them that the world is round, or had he used his supernormal powers to demonstrate it, they would not have believed him, thus hindering his ability to gain their confidence in teachings of far greater importance for the path to liberation from suffering and rebirth. Thus, the Buddha's description of a flat world, with a central mountain surrounded by a great ocean with four island continents, was a provisional teaching, set forth by the Buddha in keeping with the interests and capacities of his audience, adapted to the historical moment.

Gendun Chopel is therefore able to argue that the Buddha taught that the world is flat although in fact the world is round. However, because the Buddha knew that the world is round (and simply did not say so), the omniscience of the Buddha is preserved. This is an approach that in some ways is very traditional, despite its claim that the Buddha, renowned for never teaching with "a closed fist," knew something that he never taught: that the world is round.

The doctrine of *upāya*, of the Buddha's skillful method, is often portrayed as a sign of the Buddha's compassion, adapting his teachings to the needs of his audience, just as a doctor prescribes different cures for different maladies. All are teachings of the Buddha. The same doctrine, however, has a long history in Buddhist polemics, dating back at least to the rise of the Mahāyāna, where earlier teachings could be identified as provisional and subsumed into a new hierarchy. The early teaching is conceded as necessary for its time and place, but it is deposed to a lesser rank; the old becomes provisional, the new is definitive. The old is that which the Buddha offered to those of limited capacity, until they are prepared to receive the new teaching. For Gendun Chopel the old teaching is that the world is flat. The new teaching is that the world is round. And the Buddha knew that all along.

He concludes his brief essay with the statement, "If all of us would believe in this world that we see with our eyes rather than that world that we see through letters, it would be good." At first sight, this appears to be a call to experience, to turn one's gaze away from musty tomes and see the world as it is. Gendun Chopel himself had left the monastery and gone out into the world, and there he had learned the true shape of the world, which he now relays back to his homeland. But as he concedes earlier in the essay, the world appears to be flat and the texts describe it as such. There would seem then to be no dissonance between the world seen with the eyes and the world seen through letters. He must have a different kind of seeing in mind. For this, we must turn to his more sustained discussion of science, which is considered in chapter 3.

Another Tibetan who has taken up the question of the shape of the earth is the most prominent of Tibetans, the Fourteenth Dalai Lama. He has taken a different view, setting aside issues of the Buddha's skillful methods to conclude that the Buddha was wrong about the shape of the world. This need not shake one's faith in the Buddha, because he was right about more important things, such as the four noble truths. The Dalai Lama thus has simply rejected the traditional cosmology, writing, "The purpose of the Buddha coming to this world was not to measure the circumference of the world and the distance between the earth and the moon, but rather to teach the Dharma, to liberate sentient beings, to relieve sentient beings of their sufferings."[39] This is in keeping with the Dalai Lama's position that the relevance of science for Buddhism is confined to the first two of the four noble truths: the truth of suffering and the truth of the origin of suffering. Science has nothing to contribute on the last two truths: the truth of the cessation of suffering (called nirvāṇa) and the truth of the path to that cessation.

To concede that the Buddha might have been wrong about anything, however, is a dangerous concession to make, for it raises the question of what else he might have been wrong about, and of the criteria by which the Buddha's being right or wrong might be judged. As we have already seen, in 1863 the Christian missionary Robert Spence Hardy regarded the Buddha's error about the Mount Meru cosmology as a fatal flaw. It is useful, therefore, to pause briefly to consider the content of the Buddha's enlightenment.

In the earliest of the Pāli biographies of the Buddha, the "Account of Origins" (*Nidānakathā*), dating from the fifth century CE (and thus some eight centuries after the death of the Buddha), we find this description of the Buddha's enlightenment:

> While the sun was still shining above, the Great Being thus dispersed Māra's army; being honoured with the offerings in the form of the young leaves from the Bodhi tree falling on his robe, as though with shoots of red coral, he entered into the knowledge of previous existences in the first watch of the night; in the second watch he purified his divine eye; and in the final watch gained an insight into the knowledge of the interdependent causal origins. As he continued to reflect on the nature of the causal antecedents which consist of twelve constituents, in their direct and inverse relations in progressive and regressive evolution, the ten thousand world systems quaked twelve times up to the very limits of the ocean. When the Great Being gained penetrative insight into omniscient knowledge at dawn, making the ten thousand world systems resound, the entire ten thousand worlds assumed a festive garb.[40]

There is much that could be said about this passage. In brief, however, it recounts that after dispensing with the armies of Māra, the Buddhist demon of death and desire, Prince Siddhārtha meditated all night beneath the Bodhi tree. During the first watch of the night (the period between 6:00 PM and 10:00 PM), he had a vision of all of his past lives. In the second watch of the night (the period from 10:00 PM to 2:00 AM), he gained the "divine eye," a power variously defined, but in this context said to be the vision of how "beings pass away and come into being according to their deeds," that is, insight into the operation of the law of karma. It is noteworthy that neither of these visions constituted his enlightenment, nor are they said to be unique to the Buddha or even to

Buddhists; they are described as the by-products of states of deep meditation attainable by any yogin. It was in the third watch of the night (the period between 2:00 AM and 6:00 AM) that the prince became the Buddha. He contemplated causation in detail, examining the twelve links of the chain of dependent origin (*pratītyasamutpāda*), beginning with ignorance and ending with aging and death, forward and backward. As a result of his understanding, at dawn he "gained penetrative insight into omniscient knowledge" and at that point became the Buddha. But what is this omniscient knowledge, literally here, "the knowledge that knows everything" (*sabbaññutāñāṇa*)? Let us consider briefly the question of the Buddha's omniscience, a question upon which the various Buddhist traditions are not univocal.

In the Pāli scriptures, the nature of the Buddha's knowledge is variously described. In some cases, the knowledge of all, or omniscience, is glossed as knowledge of the dharma. This is not clarifying, since *dharma* is notoriously untranslatable. Here, however, it seems to mean "the truth" or "the nature of things." Elsewhere, the Buddha explains that he is endowed with three knowledges: the knowledge of his own past lives, the divine eye (described above), and the knowledge that he is free of all defilements. In another text, he claims knowledge of the past and the present, but not the future, apart from knowledge of the fact that when he dies he will not be reborn again. The Buddha denies that anyone can know everything in the universe at the same time; later commentators took this to imply that the Buddha can know—with direct perception and not merely by inference—anything to which he turns his mind. But this raised a problem: to know everything in the universe, one by one, would require a very long time. The accepted view on the topic in the Theravāda came to be that the Buddha could know everything in the universe that has existed, exists, or will exist (that is, he can know the past, present, and future); that he can know these things collectively as well as individually; and that he can know them simultaneously or sequentially. Those with further questions should recall that the Buddha said, "O monks, the range of objects of the Buddha's knowledge transcends all thoughts on the subject; whoever indulges in thinking about it will only suffer mental aberration and distress."[41]

The Buddha is omniscient in the Mahāyāna as well. He, and all buddhas, are said to be endowed with *sarvākarajñāna*, literally, "the knowledge of all aspects." The nature of his omniscience is described in detail, as is so often the case in Buddhism, in the form of lists. Thus, in the Mahāyāna, the Buddha is said to be endowed with two types of knowledge. The first, *yāvadbhāvikajñāna*, might be translated as the "knowledge of the multiplicities" or the "knowledge of the varieties." This is the Buddha's knowledge of each of the phenomena of the universe in its specificity. The second is the *yathāvadbhāvikajñāna*, which might be translated as "the knowledge of the mode," that is, the Buddha's understanding of the single mode of being of the universe. Only a buddha possesses these two knowledges and possesses them simultaneously. He is able to do so because only a buddha has abandoned the two obstructions. The first are the afflictive obstructions (*kleśāvaraṇa*), which include desire, hatred, and ignorance. He has also abandoned the second type of obstructions, the *jñeyāvaraṇa*, literally, "the obstructions to objects of knowledge," subtle predispositions that prevent the simultaneous perception of objects and their true nature. The Buddha thus is able to perceive all of the various phenomena of the universe as well as their final nature. This is referred to as the simultaneous knowledge of the two truths, the conventional truth (*saṃvṛtisatya*) and the ultimate truth (*paramārthasatya*).

In addition, the Buddha is said to be endowed with five wisdoms. These are (1) the *dharmadhātujñāna*, or "wisdom of the sphere of reality"; (2) the *ādarśajñāna*, or "mirrorlike wisdom," which reflects reality exactly as it is; (3) the *pratyavekṣaṇajñāna*, perhaps translated as "discriminating wisdom" or the "wisdom of specific understanding," direct understanding of the general and specific characteristics of all the phenomena in the universe; (4) the *samatājñāna*, or the "wisdom of equality," in which no ultimate difference is perceived between saṃsāra and nirvāṇa in terms of the former having a nature of evil and the latter having a nature of goodness; and (5) the *kṛtyānuṣṭhānajñāna*, or the "wisdom of accomplishing activities," the wisdom that works for the welfare of all beings and serves as the cause for the various emanations of the Buddha.

There are also the eighteen *āveṇikadharma*, the "unshared attributes" possessed only by the Buddha. Here we might cite the Dalai Lama's own description of them, drawn from his first book on Buddhism, *Opening the Eye of New Awareness* (*Blo gsar mig 'byed*), published in Tibetan in 1963. The Buddha is endowed with the attributes of (1) not being mistaken, such as being frightened of thieves, tigers, and the like when traveling in towns, cities, forests, et cetera; (2) not having uncontrolled speech, such as letting out loud cries upon losing one's way or bursting out laughing due to the influence of predispositions; (3) not having memory lapses, such as letting an activity slip due to forgetting it, or being late for an activity; (4) not having a mind that is not in *samādhi* (one-pointed concentration) on the meaning of emptiness at all times, whether in or out of meditative absorption; (5) not having the discrimination of difference, which is the conception that saṃsāra is inherently established as unfavorable and that nirvāṇa is inherently established as peaceful; (6) not having the indifference that neglects the welfare of sentient beings upon not individually analyzing such things as the appropriate time for taming them; (7) the uninterrupted arising of the aspiration for love and compassion, bringing about the welfare of sentient beings; (8) the effort that is the enthusiasm for going to buddha lands that surpass the number of grains of sand on the banks of the Ganges for the sake of even one sentient being; (9) the constant mindfulness never to forget the styles of mental behavior of all sentient beings as well as the methods for taming them; (10) the *samādhi* set in equipoise on the suchness of phenomena; (11) the wisdom that knows how to teach appropriately the 84,000 bundles of doctrine as antidotes to the afflictive styles of behavior of disciples; (12) the nondeterioration of the liberation that is the state of having abandoned all obstructions exhaustively; (13) the exalted physical activities, such as emitting light, the four postures, et cetera; (14) the exalted verbal activities of teaching in accordance with the inclinations of sentient beings; (15) the exalted mental activities endowed with great love and compassion; (16) unimpeded great knowledge of all objects of knowledge of the past; (17) the unimpeded great knowledge of all objects of knowledge of the present; and (18) the unimpeded great knowledge of all objects of knowledge of the future.[42]

Without enumerating yet other qualities of the Buddha, such as the ten powers, the four fearlessnesses, the four knowledges, the eight liberations, the four specific knowledges, the four purifications, and the six superknowledges, it is possible to describe the Buddha in the following terms. He has complete recollection of the past, including each of his own past lives as well as those of all sentient beings (the past lives of each being are said to be infinite in number). He has full knowledge of the present in the sense that he is aware, or has the capacity to be aware, of all events occurring in all realms of multiple universes. He has full knowledge of everything that will occur in the future and is able to predict the precise circumstances under which various persons will become buddhas. He has direct and complete insight into the most profound nature of reality at all times, such that he does not need to enter into meditation in order to see it directly. He is fully aware of the contents of the minds of all beings in the universe, and thus is able to teach them in accordance with their individual needs and dispositions. He is endowed with a perfect comportment, always displaying complete dignity in both his physical movements and his speech (according to some accounts, the neck bone of all buddhas is solid such that they never turn their head, always facing forward). He is perfectly enlightened and hence omniscient.

This view of the Buddha is widely shared across the Buddhist traditions. However, at least for one important Indian Buddhist thinker, it is not his knowledge of the facts of the world that makes him worthy of respect and devotion. In his *Pramāṇavārttika* (Commentary on Valid Knowledge), the seventh-century logician Dharmakīrti takes up the question of what makes a teacher (in this case, the Buddha) valid and authoritative. Without taking a position on whether the Buddha was omniscient or not, he makes it clear that knowledge of mundane facts or the ability to see things beyond the range of normal vision is not relevant.

> Thus, one should investigate [if a teacher]
> Has the knowledge of what should be practiced.
> Whether he knows the number of insects
> Is not of any use to us.
>
> We seek one who is valid,
> Knowing what to adopt and discard
> As well as the method to do so,
> Not one who knows everything.
>
> Whether or not he can see what is distant,
> He should see the reality we seek.
> If seeing what is distant makes one valid
> Then we should honor vultures here.[43]

This trenchant point is similar to the Dalai Lama's statement above that the Buddha's purpose was not to measure the distance between the earth and the moon but to teach the path to liberation. It is important to note, however, that unlike the Dalai Lama, Dharmakīrti does not concede that the Buddha was wrong about anything.

Perhaps in consideration of this problem, in his more recent writings the Dalai Lama does not ascribe the Mount Meru cosmology to the Buddha, but rather to the fourth-century Indian master Vasubandhu. It is Vasubandhu's *Abhidharmakośa* that forms one of the five books (*gzhung lnga*) of monastic curriculum in the Dalai Lama's sect, the Geluk, and it was from the *Abhidharmakośa* that the Dalai Lama learned the traditional cosmology during his education in Tibet. In his recent *The Universe in a Single Atom*, he writes: "By the age of twenty, when I began my systematic study of the texts that discuss Abhidharma cosmology, I knew that the world was round, had looked at photographic images of volcanic craters on the surface of the moon in magazines, and had some inkling of the orbital rotation of the earth and moon around the sun. So I must admit, when I was studying Vasubandhu's classic presentation of the Abhidharma cosmological system, it did not much appeal to me."[44]

Thus, like Gendun Chopel in the 1930s, for the Dalai Lama in the 1950s it was the exposure to the discoveries of modern science—or perhaps more accurately, their representation in popular media—that provided a challenge to the traditional Buddhist cosmology. It is not sufficient, however, simply to dismiss the traditional views; those views must be interpreted in order to find a place for them within a larger, and less problematic, Buddhist schema. As we have seen, Gendun Chopel made a traditional appeal to the doctrine of *upāya*, the Buddha's skillful methods, compassionately teaching that which accorded with the capacities and interests of his audience, even when he himself knew it to be untrue. The Dalai Lama seems to take a different tack.

> These sizes, distances, and so forth are flatly contradicted by the empirical evidence of modern astronomy. There is a dictum in Buddhist philosophy that to uphold a tenet that contradicts reason is to undermine one's credibility; to contradict empirical evidence is still a greater fallacy. So it is hard to take the Abhidharma cosmology literally. Indeed, even without recourse to modern science, there is a sufficient range of contradictory models for cosmology within Buddhist thought for one to question the literal truth of any particular version. My own view is that Buddhism must abandon many aspects of the Abhidharma cosmology.

> To what extent Vasubandhu himself believed in the Abhidharma worldview is open to question. He was presenting systematically the variety of cosmological speculations that were then current in India. Strictly speaking, the description of the cosmos and its origins—which the Buddhist texts refer to as the "container"—is secondary to the account of the nature and origins of sentient beings, who are "contained."[45]

Vasubandhu is one of the most revered Indian masters in Tibetan Buddhism, venerated as one of the "six ornaments of the world," along with his brother Asaṅga, as well as Nāgārjuna and Āryadeva, and Dignāga and Dharmakīrti. Although an early proponent of the Hīnayāna, he was converted to the Mahāyāna by his brother, and went on to compose important works on Yogācāra philosophy. In the Dalai Lama's own sect, the Yogācāra is considered philosophically inferior to the Madhyamaka. But it is also said that although Asaṅga taught Yogācāra, his own view was Madhyamaka.[46]

The Dalai Lama's comment above, that the extent to which Vasubandhu himself believed in the Abhidharma worldview is open to question, can thus be read in several ways. Later in his life, Vasubandhu did indeed change his philosophical position, although the Mount Meru cosmology is held in common by the schools of Indian Buddhism, both Hīnayāna and Mahāyāna. The more likely reading is that the Dalai Lama is making a claim for Vasubandhu similar to the one that Gendun Chopel made for the Buddha: he knew that the earth is not flat, but he said that it was, in keeping with the conventions of the time.

This claim raises a number of fascinating questions about when and where the words of the Buddha are to be taken at face value, and when they should be declared merely provisional, questions that the tradition has struggled with for two millennia, questions that have particular significance in the domain of Buddhism and Science.

But it is clear that a number of Buddhist thinkers, across the Buddhist world during the past two centuries, have been willing to abandon the traditional cosmology, seeing it, in one sense or another, as inessential to Buddhism. Other elements of the tradition are defended more steadfastly, as we will see in chapter 3. What is not often noted in these discussions is the relationship between Buddhist cosmology and those Buddhist doctrines that have tended to be deemed more important during the history of the discourse of Buddhism and Science. As the Dalai

Lama notes above, "the description of the cosmos and its origins—which the Buddhist texts refer to as the 'container'—is secondary to the account of the nature and origins of sentient beings, who are 'contained.'" But where do the "contained," the beings who populate the realm of rebirth, abide, if not in the "container"?

Mount Meru is not simply a barren peak, like Everest. The ocean that surrounds the mountain and the lower reaches of its four faces are the abodes of all manner of demigods, one of the six types of beings who populate the Buddhist universe and as which all beings have been reborn. Its upper reaches are the kingdoms of the monarchs of the four directions, who survey the world from their palaces on the mountain. Among these is Vaiśrāvaṇa, the king of the north and the god of wealth, and, as such, one of more popular gods in the Buddhist pantheon. On the flat summit of Mount Meru is located the Heaven of the Thirty-Three, among the most commonly mentioned of the Buddhist heavens. It is the abode of Śakra (Indra), who converses so frequently with the Buddha and who appears in so many Buddhist sūtras. It was in the Heaven of the Thirty-Three that the Buddha spent the rains retreat in the seventh year after his enlightenment, when he taught the assembled gods, including his mother. She had died shortly after his birth and been reborn as a male god. In order that she not be deprived of the teachings of the dharma, the Buddha traveled to the Heaven of the Thirty-Three to teach her there. He taught her the Abhidharma, one of the "three baskets" that constitute the Buddhist canon, the section that contains central expositions of the nature of consciousness and the operation of rebirth. The Buddha's descent from the mountain back to the continent of Jambudvīpa is one of the most celebrated of his deeds. His return from the mountain on a divine ladder made of gold, silver, and jewels, met at the bottom by the monk Śāriputra and the (disguised) nun Utpalavarṇā, is among the most famous scenes in Buddhist art. The place of his descent, Sāṃkāśya, is one of the sacred sites of the tradition because all the buddhas of the past have descended from Mount Meru there.

Some Buddhist thinkers wanted to keep Mount Meru on earth, yet beyond the reach of explorers. Others placed it in outer space. Still others placed it in the category of the nonexistent, consigned to the realm of myth, without fearing that any harm has been done to the dharma. That this would occur was predicted with remarkable prescience almost a century and a half ago, not by a Buddhist monk, but by a Christian missionary. In a lecture delivered at the Union Church in Hong Kong in the winter of 1871, Ernst Eitel declared:

> As to the popular literature of Buddhism and its absurdities, we might as well collect those little pamphlets on dreams, on sorcery, on lucky and unlucky days, on the lives and miracles of the saints, which circulate among Roman Catholic peasants—but would that give us a true picture of Roman Catholicism? Thus it is with Buddhism.
>
> Those crude, childish and absurd notions concerning the universe and physical science do not constitute Buddhism. This great religion, imperfect and false as it is to a great extent, does not stand or fall with such absurdities. They are merely accidental, unimportant outworks, which may fall by the advance of knowledge, which may be rased to the ground by the progress of civilisation, and yet the Buddhist fortress may remain as strong, as impregnable, as before. A Buddhist may adopt all the results of modern science, he may become a follower of Newton, a disciple of Darwin, and yet remain a Buddhist.[47]

Yet once the process of demythologizing begins, once the process of deciding between the essential and the inessential is under way, it is often difficult to know where to stop. The question, then, is which Buddhist doctrines can be eliminated while allowing Buddhism to remain Buddhism. Can there be Buddhism without Mount Meru? Can you play chess without the queen? Mount Meru—with its four faces of gold, silver, lapis, and crystal—is a slippery slope.

In 1977 I was discussing the traditional cosmology with a prominent Tibetan lama. I asked him, somewhat more politely than the Christian missionaries in Sri Lanka, why it was that Mount Meru had not been discovered. We were speaking Tibetan, and his answer could be translated in two ways. The first would be, "If one has pure karma, one can see it." The second would be, "If you had pure karma, you could see it."

BUDDHISM AND THE SCIENCE OF RACE

On October 27, 1937, the following letter, written in perfect German, arrived in Berlin:

> Abbot T'ai-hsü, Monastery Ta-lin Szu
> Kuling via Kiu-kiang (Kiangsi)
> China
>
> Kuling, 11 August 1937
>
> To the Leader of the German People,
> Mr Adolf Hitler.
>
> The scientific civilization of our time is borne by the Aryan race, but the religious culture of the past has its culmination in Buddhism, whose founder, Buddha Shakyamuni, was also of Aryan origin.
>
> People in Europe and America today are not happy, obviously because their lives are ordered by science alone, which offers no answers to questions concerning religious issues. They are in need of religion. Now, most

religions stand in contradiction to natural science; only Buddhism has fully absorbed its insights, indeed surpassing them. Thus Buddhism is destined to become the religion of the peoples of Europe and America.

Buddhism has recognized the fundamental truth that there are four primary virtues that man must possess in order to achieve perfection: compassion (for the needs of one's neighbour), conformity (to the social hierarchy), good works (for improvement), and courage (to break down opposition). The peoples in India and China possess the first two of these virtues, but they do not have the last two in sufficient measure. Thus their personality is incomplete and the full extent of the blessing of Buddhism cannot yet reveal itself to them.

I believe that the Germanic people, now united under their Führer, have wondrously developed three characteristics: knowledge, conformity, and courage. Thus only the Buddhist religion, in which these three characteristics are primary virtues, can be the religion of the Germanic people. And only that most excellent scion of ancient Aryan stock, Shakyamuni, the Holy, can be the religious leader of the Germanic people, that most excellent scion of ancient Aryan stock.

If the Führer desires to study the Buddhist religion, which can become so important for Europe and America today and for the Germanic people, I request that he write to me and I will gladly answer to the best of my knowledge.
I wish your government undaunted stability!

[several seals]
Leader of the Buddhists in China
signed T'ai-hsü[1]

On April 26 of that same year, German Junker and Heinkel bombers had destroyed the Basque town of Guernica. On July 16, a concentration camp was opened on the Ettersberg plateau overlooking the Weimar River, a region closely associated with Goethe. The camp was called "Forest of Beech Trees," or Buchenwald. Also in 1937, Jews were ordered to wear a yellow Star of David on their clothing and were barred from entering parks, theaters, health resorts, and public institutions.

The chronological proximity of these events to the arrival of the letter does not suggest that Taixu (as "T'ai-hsü" is rendered in the modern system) was somehow in sympathy with Hitler's policies; it is highly unlikely that he was aware of them. And he was not the only Buddhist leader to write to the German dictator; the regent of the young Fourteenth Dalai Lama, Reting Rinpoche, also did so, in 1939. The more important event of 1937, however, occurred not in Hitler's Germany but in Taixu's China, when Japanese forces invaded on July 7. Beijing fell three weeks later. It appears likely that it was this event, more than any particular attraction to Nazi ideology, that motivated Taixu's letter, written two months after the Japanese invasion had begun.

Taixu opens the letter with an appeal to the term *Aryan*, a term central to both Buddhism and Nazism. He next declares that Europe has science but it does not have religion, or, at least, it does not have the right religion. Europe requires a religion that is compatible with science, a scientific religion, and that religion is Buddhism. Furthermore, the most appropriate Europeans to adopt Buddhism are the Germans. The Aryan people are naturally suited to the Aryan religion. The purpose of this conversion, or perhaps return, however, is not simply to restore the natural order. The Indians and Chinese (and it is noteworthy that he does not mention the Japanese), custodians of Buddhism for centuries, are constitutionally incapable of fully appreciating it. The solution, then, was for Europe to convert to Buddhism and then missionize Asia. Taixu's bizarre dream, or so it seems, was for European Buddhist missionaries to proselytize the elites of China and India. He, an Asian Buddhist, would spread the dharma in Europe (or at least answer Hitler's questions) in order that German Buddhists could spread the dharma among Asians. There was likely an urgency to Taixu's call; three hundred thousand Chinese would die at the hands of the Japanese Imperial Army in the Nanjing Massacre two months after his letter arrived in Berlin.

Regardless of the motivation for Taixu's letter, it is clear that by 1937, a different European science had entered the discourse of Buddhism and Science: the science of race. This chapter will trace the long route by which it arrived.

• • •

Seven weeks after his enlightenment, the Buddha sat reflecting upon the profundity of what he had understood. So profound was it that he was reluctant to speak of it to others. At that moment, he was visited by the god Brahmā Sahampatī, who implored him to teach. Moved by the deity's plea, the Buddha surveyed the world with his divine eye and saw that there were two kinds of beings in the world: those with little dust in their eyes and those with much dust in their eyes, those with keen faculties and those with dull faculties, those with good qualities and those with bad qualities, those easy to teach and those difficult to teach.[2] He then set out to find someone worthy of his teaching. Noting that his two former teachers of meditation had recently died, he selected the "group of five," his former comrades in asceticism. He walked to where they were residing, the Deer Park outside Banaras, and after proclaiming his enlightenment to them, he preached to them or, in the language of the tradition, he turned the wheel of the dharma. He first described the two extremes of self-indulgence and asceticism and the middle path between them. He then said, "idaṃ kho pana bhikkhave dukkhaṃ ariyasaccaṃ," usually translated as "Now this, O monks, is the noble truth of suffering."

The Buddha's attitude toward the so-called caste system of ancient India has obsessed the study of Buddhism for almost two centuries.[3] It has often been stated in the most straightforward terms, with the opposing positions clearly marked. On the one side is the Hindu, or Brahmanical, tradition, which upholds the fourfold division of society into a hierarchy of the twice-born—the brahmans, *kṣatriya*s, and *vaiśya*s—who have access to the sacred Veda, and the fourth caste, the *śūdra*s, who do not. This prohibition is most graphically stated in an early law book called the *Gautama Dharmasūtra*, which states (at 12.4), "If a *śūdra* intentionally listens to the Veda, then his ears should be filled with [molten] lead and lac; if he utters the Veda, then his tongue should be cut off; if he has mastered the Veda, his body should be cut to pieces." This passage is quoted approvingly by later figures, including the famous philosopher Śaṃkara himself. Similar, although less graphic, statements are found in the more famous *Manavadharmaśāstra*, or "Laws of Manu." For example, the eightieth and eighty-first verses of the fourth chapter state, "He must never give a Śūdra advice, leftovers, or anything offered to the gods; teach him the Law; or prescribe an observance to him. Whoever teaches him the Law or whoever prescribes an observance to him will plunge along with him into that darkness called [the hell of] Asaṃvṛta."[4] As this particular presentation of the caste system so commonly reminds us, the division of society into these four groups is sanctioned by the Vedas; the *Puruṣaśukta* of the Rig Veda explains that the four castes emerged at the creation of the universe from the sacrifice of a primordial being, with the brahmans emerging from his mouth, the kṣatriyas from his arms, the vaiśyas from his legs, and the śūdras from his feet. Outside the fourfold division are groups referred to generically as "the fifth" (*pañcama*), who engage in a variety of polluted and polluting professions. These are the so-called untouchables.

According to the standard presentation, on the other side stands the Buddha, who claimed that anyone, regardless of caste, could follow the path to nirvāṇa, and who admitted members of all four castes into his order of monks. The Buddha repeatedly questioned the brahman's claim to superiority due to birth, noting that because their wives menstruate, become pregnant, give birth, and nurse babies, brahmans are born from the wombs of women and not from the mouth of the god Brahmā; their claim to divine origin is therefore just a lie.[5] The Buddha thus dismissed caste hierarchy entirely, it is said, shifting the valuation of the person from blood and birth to virtue and morality. The Buddha was therefore not merely a teacher of the path to liberation from the world of suffering, but was also a reformer of that world.[6] His efforts, however, were short-lived; Buddhism eventually disappeared from India, and the caste system survived. As the eminent Victorian translator and interpreter of Buddhism Thomas W. Rhys Davids (1843–1922) wrote, "Had the Buddha's view on the whole matter won the day the evolution of social grades and distinctions would have gone on in India on lines similar to those followed in the west, and the caste system would never have been built up."[7]

This Victorian view of the Buddha's attitude toward caste achieved the status of common knowledge in both Europe and India, promoted especially by the great untouchable convert to Buddhism Dr. Bhimrao Ramji Ambedkar (1891–1956) and his followers. Like so many representations of Buddhism deriving from this period, however, the situation is a good deal more complicated, as was duly recognized by a number of the leading scholars of the day, including Hermann Oldenberg, Émile Senart, and Max Weber himself.

Entire volumes have been written, and continue to be written, on Buddhist attitudes toward caste, and only the barest disconnected discourse on current scholarly opinion can be provided here. The Buddha himself was a member of the kṣatriya, or warrior caste, and when he made reference to the fourfold caste division, as he would most commonly do when his interlocutor was a brahman, he would not list the four in the standard order of brahman, kṣatriya, vaiśya, śūdra, but instead would place the kṣatriyas first. In other contexts, he commonly spoke of only three groups, which do not appear in a hierarchy: the kṣatriya, brahman, and *gṛhapati*, or householder.[8] He explained at one point that kṣatriyas are superior to brahmans because they do not allow into their caste children born of a union in which only one parent is of the higher caste, whereas the brahmans do.[9] In describing the buddhas of the past, he said that a buddha is never born into a vaiśya or śūdra family, but only as the son of brahmans or kṣatriyas, depending on which of the two is more highly honored by society at the time of his birth—the implication being, of course, that he himself chose to be born as a kṣatriya because it was the superior caste of his day.[10] Apart from such statements, the Buddha tended to emphasize occupation over caste status, and reserved the term *brahman* for the most advanced yogins. For example, in the *Vāseṭṭha Sutta* of the *Sutta Nipāta* (considered an early work), he says, "Whoever among men makes his living keeping cows, thus know, Vāseṭṭha, he is a farmer, not a brahman. . . . Without anger, possessing vows and virtuous conduct, free from haughtiness, tamed, having his last body, him I call a brahman."[11]

Although members of all castes joined the order of monks and nuns, the Buddha drew his followers, both monastic and lay, largely from the brahman and kṣatriya castes, if we can draw conclusions from the canon. Of the 105 monks and nuns whose caste is mentioned in the Pāli *tipiṭaka*, some 60 percent are from the two upper castes, with brahmans forming the largest group. There are indeed members of the "low class" (*nīca kula*), but only eight: two barbers, one potter, one fisherman, one vulture trainer, one son of a slave, one elephant trainer, and one actor.[12] Of these, only one achieved prominence. This was Upāli, the barber of the Buddha's own Śākya clan, who would become renowned as a master of the monastic code. But even as a monk, he was once mocked by some angry nuns, who called him a *mala majjano*, "one who shampoos the hair of others."[13] A postulant to the monastic order is typically described as *kulaputra*, a "son of good family."

The Buddha regarded the state of the renunciant or *śramaṇa* as beyond the category, and thus restrictions, of caste. Such a person had gone forth to the homeless life and therefore no longer had a caste designation. In this regard, however, the Buddha seems only to reflect the ethos of his milieu: Jains and Ājīvikas made the same claim. These groups were united in their rejection of the Brahmanical claim to sacred knowledge and to the primacy of their cult of sacrifice, which together formed the basis of the brahmans' claim to preeminence. Thus, the caste of a person who joined the order of monks and nuns was irrelevant because caste status and duties were renounced upon ordination. When the Buddha's father protested that it was unseemly that his son, a member of the kṣatriya lineage, should beg for his food, the Buddha replied that it was his father who was of the royal lineage; his lineage was that of the buddhas.[14] Indeed, one who has joined the saṅgha is called in Pāli *vevaṇṇiyanti*, "one without caste," and is compared to a river that has lost its identity when it flows into the sea. The Buddha, nonetheless, is praised for being of pure paternal and maternal descent, going back seven generations.[15]

In his account of the origin of the universe in the *Aggañña Sutta*, the Buddha explains that the first caste to be established was the kṣatriya, when the first king was selected by popular demand because he was "the handsomest, best-looking, the most pleasant and capable."[16] Sometime after that a group decided to renounce unethical behavior, live in the forest in huts made of leaves, and meditate all day, only going into villages to beg for food (very much like Buddhist monks). These were the first brahmans. However, a group of these renunciants stopped meditating; they moved into the village and began to compile the Vedas. The Buddha describes these people as the inferior brahmans, noting that in recent times they had come to be regarded as superior.[17] It is this group, the householding class of priests, the performers of sacrifice, who are clearly the Buddha's chief antagonists and his chief competitors; and it is this group, and only this group, that he consistently criticizes as prideful and avaricious.[18]

The Buddha speaks (as do Jain texts) of the state of being a true brahman as something gained not through bloodline and birth, but through virtue. As is so often the case in Buddhism, it all comes down to karma: it is one's deeds in a past lifetime that cause one to be reborn in a high caste or a low caste in the present lifetime, and it is one's deeds in the present lifetime that will cause one to be reborn in a high caste or a low caste in a future lifetime.

To summarize, then, the Buddha lived in a society in which caste distinctions largely defined the social order. Like other *śramaṇa* teachers of the day, he regarded the state of the renunciant as beyond the boundaries of caste. At the same time, he was critical of brahman priests whose status depended on caste and who, it should be noted, were his foremost competitors for patronage. He sought to convince his followers that the giving of gifts to Buddhist monks was a more efficacious form of religious practice than paying brahman priests to perform sacrifices. There is no evidence that the Buddha sought to "reform" or destroy what has been called the caste system. Like his contemporaries, he saw caste and clan and family to be definitive constituents of the person, as one of the most famous descriptions of his enlightenment experience testifies:

> I directed it [my mind] to the knowledge of the recollection of past lives. I recollected my manifold past lives, that is, one birth, two births, three births, four births, five births, ten births, twenty births, thirty births, forty births, fifty births, a hundred births, a thousand births, a hundred thousand births, many aeons of world-contraction, many aeons of world-expansion, and many aeons of world-contraction and expansion. "There I was so named, of such a clan, of such a caste, such was my nutriment, such was my experience of pleasure and pain, such was my life-term; and passing away from there, I reappeared elsewhere; and there too I was so named, of such a clan, of such a caste, such was my nutriment, such was my experience of pleasure and pain, such was my life-term; and passing away there, I appeared here." Thus with their aspects and particulars I recollected my manifold past lives.[19]

Buddhist texts would speak repeatedly of clan and family and occupation and craft, but tended to speak in terms not of a fourfold hierarchy, but of a binary distinction, between high (*ukkaṭṭa*) and low (*hīna*). Farming, raising cattle, and trade were high; guarding a storeroom and making baskets were low. Writing, accounting, and royal service were high; pottery, weaving, and leatherwork were low. These would persist throughout the history of Buddhism in India, where on the one hand, Buddhist logicians would continue to attack all Brahmanical claims to an ontological status for caste,[20] while on the other, even in the purportedly more universal Mahāyāna, terms like *kulaputra*, "son of good family," and *gotra*, "lineage," would abound. Even in this "metaphorical" sense, scholastic treatises would debate whether some beings are, as it were, genetically destined to one or another of the vehicles to enlightenment, and whether some beings were doomed to none.[21] As has been noted, "Purity of blood and unblemished lineage was extremely important to the Buddhists."[22]

But that is another story. We must turn now to another binary and how it was deployed in the construction of Buddhism as a world religion well suited to modernity. The two terms, in Sanskrit, are *āryan* and *anāryan*, "āryan" and "not āryan." The term *āryan* appears in the Buddha's first sermon, where he speaks of the *ariyamāgga* and *ariyasacca*. These terms have long been rendered as "noble path" and "noble truth." What they mean in the original Pāli is something of a grammatical conundrum.[23] But for the long tradition of commentary at least, it seems clear that this famous translation is inaccurate: it is not the truths that are noble, but rather those who understand them. Suffering, origin, cessation, and path are truths, or facts, only for those who are somehow "noble." For all others, they are not true. As Vasubandhu states in the *Abhidharmakośabhāṣya*, "They are the truth for the āryans, truths of the āryans; this is why they are called *āryasatya*."[24] "Noble" is perhaps an appropriate translation of *ārya* because it carries the twin connotations of superiority by birth and superiority by character. And again we find ourselves confronted with another apparent opposition, this time between the pre-Buddhist (or non-Buddhist) meaning of a term, and the Buddha's reinterpretation of it. In this case, however, the term is even more charged than the terms *caste* or *brahman*, although it is intimately related to their history.

As that history has long been told, sometime in the fourth millennium BCE a nomadic people, skilled in the use of horse and chariot, invaded the Indian Subcontinent from what is today Iran and Afghanistan. They swept through the Indus River valley, defeating the remnants of a once-great urban culture, moving eventually east to settle in the Ganges basin. They called themselves Aryan (a term cognate to Iran and Ireland) and subjugated the local population, whom they called *dāsa* (a word variously rendered into English as "fiend," "barbarian," and "slave"). The religion of the Aryans was one of sacrifice to their thirty-three gods, and their society was organized by caste, with the first three castes composed of Aryans and the fourth and lowest caste populated by the conquered peoples. Their language, which nineteenth-century philologists would call Indo-Aryan, Indo-German, or Indo-European, was related to Greek and Latin and German and English, because the speakers of these languages were originally one people.

This story, which has been told and retold in classrooms and textbooks for more than a century, has in recent years been called seriously into question, with some scholars rejecting this long-held theory of the Aryan invasion and conquest of northern India on the basis of linguistic and archaeological evidence.[25]

Regardless of the geographical origins of the bloodline called Aryan, the Buddha, born into a *kṣatriya* clan, shared this blood. He was an Aryan. Yet he gave the term an additional meaning. One day the Buddha encountered a group of monks near the north gate of Sāvatthi. He stopped to converse with them and asked them their names. There was a fisherman standing nearby who said that his name was Ariya. The Buddha observed that his name did not match his occupation, because a true *ariya* (Pāli for *ārya*) did not injure living beings. At the end of the Buddha's discourse, the fisherman (who had presumably spent a good deal of time standing in streams), became a *sotāpanna*, a stream enterer, one who has reached the initial stage of enlightenment.[26]

This story, found in the commentary to the *Dhammapada*, itself requires some commentary. On one level, it seems just a further confirmation of the Buddha's noble substitution of ethics for blood. The fisherman's work is ignoble, and so, if he wishes to rectify his name, he should abandon it. But regardless of the complications surrounding the term *āryan* in ancient India, it also carried connotations of class and ethnicity. And from this perspective, the dissonance between the fisherman's name and his occupation is that he is called *āryan* but he is performing the work of a non-*āryan*. We know that the ancient Indian law books carried sanctions for those who performed tasks inappropriate to their caste responsibilities, with the heaviest penalties reserved for those who engaged, or attempted to engage, in deeds deemed above their station. This is not the case in the Buddhist story. The fisherman presumably inherited his profession and is thus a low-caste person by birth. How he was ever named Ariya is not explained. There is clearly a disjunction, however, and this is what the Buddha seeks to remedy.

The adjective *ārya* appears with great frequency in Buddhist texts, modifying nouns like truth (*satya*), dharma, person (*pudgala*), view (*dṛṣṭi*), speech (*vāca*), and path (*mārga*). When one requests permission to become a monk, one asks to go forth (in Pāli) into the *ariyadhammavinaya*, the dharma and the discipline of the *āryan*. It is important to note, however, that the Buddha appears also to accept the widely held connotation of *āryan* as referring to those who have language, or at least proper language: *āryan* in this context is contrasted with *mleccha*, a term generally translated as "barbarian."[27]

The more pertinent impression for our purposes, however, is that the Buddha, or at least the early Buddhists, appear to have sought to remove *āryan*'s connotation of an inherited superiority, and give it the meaning of an acquired superiority—a superiority, it must be said, held only by Buddhists, and eventually, only by some Buddhists. The Buddha's attitude toward the term *āryan* thus is a kind of correlate to his attitude toward caste. The person who enters the order of monks and nuns loses thereby his or her caste identity. At the same time, one gains a new identity, the identity of an *āryan*.[28]

On his alms round, the monk Aṅgulimāla is moved by the suffering of a mother and her newborn child. The Buddha recommends that Aṅgulimāla cure them by an "act of truth," a declaration whose truth has supernatural powers, in this case the power to heal. The Buddha first instructs him to say, "Sister, since I was born, I do not recall that I have ever intentionally deprived a living being of life. By this truth, may you be well and may your infant be well!" Aṅgulimāla politely points out to Buddha that this is not entirely accurate. Prior to his ordination, Aṅgulimāla had been a notorious serial killer of 999 persons; his name means "Necklace of Fingers." The Buddha therefore amends the statement to begin, "since I was born with noble birth." The phrase "noble birth" can be interpreted in a number of ways, but here it seems to mean "since I became a monk." When Aṅgulimāla speaks these words to the mother and her child, they are cured.[29]

The early texts would thus seem to suggest that anyone who entered the order thereby acquired this status. However, as the tradition developed, the title of *āryan* would become more exclusive, reserved only for the so-called eight noble persons (*ariyapuggala*), those who had achieved various levels of the path to nirvāṇa: the stream-enterer, the once-returner, the never returner, and the arhat; the least of these would never be reborn as an animal, ghost, or hell being again and was destined to achieve nirvāṇa in seven lifetimes or less. This was clearly an elite group, one which the tradition regarded as increasingly small in the centuries after the Buddha's own passage into nirvāṇa. The transformation of the fisherman Ariya is thus all the more remarkable. He becomes a true *ariya* (in the technical sense of a stream-enterer), on the spot. Such was the power of the Buddha's discourse. As the tradition developed, the rank of the *āryan* became more and more remote. According to the Mahāyāna treatises, for example, the bodhisattva only became an *āryan* on the third of the five paths, the *darśanamārga*, or path of vision. Scholiasts explain that this occurs 10^{59} aeons after taking the bodhisattva vow to achieve buddhahood for the welfare of others. In Buddhist scholastic literature, *āryan* is commonly paired with its opposite—not *anāryan*, "non-aryan," but *pṛtagjana*, a term usually translated as "common beings" but which also means simply "other people."

And now, to Europe. In 1844 the French Sanskritist Eugène Burnouf (1801–1852, whom we will encounter again in chapter 4) published *Introduction à l'histoire du Buddhisme indien*. There he writes, "We now see, if I am not mistaken, how this celebrated axiom of Oriental history, that Buddhism erased all distinction of caste, must be understood."[30] It is noteworthy that at the time that the academic study of Buddhism in Europe was in its very infancy, Buddhism's rejection of caste was already well known, a "celebrated axiom," one which, it should be noted, continues to be repeated. Yet the European travelers, colonial officials, missionaries, and scholars who propounded this axiom did so without having examined any Indian Buddhist texts (with the exception of Brian Houghton Hodgson's 1829 "A Disputation Respecting Caste by a Buddhist").

The first biography of the Buddha to be composed by a European scholar (apart from various versions of his life that appear in the accounts of missionaries and travelers) was "Leben des Budd'a nach Mongolischen Nachricht" by the German Mongolist Julius von Klaproth (1783–1835) and published in his *Asia Polyglotta* in 1823. There we read, "Buddha appeared as a reformer of the dominant religion of India. He rejected the Vedas, blood sacrifice, and the distinctions of caste. As for the rest, the philosophical principles and his doctrine are the same as those encountered in the other branches of the religion of the Hindus."[31] In Edward Upham's eccentric 1829 work, *The History and Doctrine of*

Budhism, Popularly Illustrated (the first book published in English with the word *Buddhism*, actually, *Budhism*, in its title), we read:

> The great schism which divided the Eastern world, and made the disunion irreconcilable, seems in fact to have originated in the period wherein the munis, or teachers of the Budhist doctrine, either from a reforming principle, or a love of power, or a combination of both, proceeded to have their own theories and sacred books, not explanatory of, but in direct opposition to, the Vedas; teaching their followers that they alone were true believers of the saving faith, throwing down the barriers of caste, and elevating the dogmas of the faith above the sacerdotal class, and admitting every one who felt an inward desire to the ministry and preaching of their religion.[32]

In 1835 Charles Neumann wrote that the Buddha sought "the entire subversion of the edifice of castes, and consequently at reforming the social system of the Hindus."[33]

The authors who described the Buddha as a "reformer" who rejected the caste system (and the brahman caste's practice of animal sacrifice) did so without consulting a Buddhist text composed in India. This "celebrated axiom of Oriental history" seems to derive instead from Hindu sources or, more accurately, Hindu sources as interpreted for the European community of scholars by Sir William Jones (1746–1794).

In an essay written in 1788, "On the Chronology of the Hindus," Jones examined the Hindu doctrine of the ten incarnations of the god Viṣṇu. He was particularly struck by parallels between Hindu stories of a great flood and the account in the book of Genesis; like most scholars of his day, he held to the Mosaic chronology. Seeking to establish the date of the flood, Jones attempted to establish the chronology of the incarnations, turning first to the ninth and most recent (the tenth incarnation, Kalki, is yet to appear), the Buddha. The Buddha was not only the most recent incarnation historically; his existence, and dates, had been reported to various missionaries to Asia in the previous century. Jones cites, among others, the work of the Belgian Jesuit Phillipe Couplet (1628–1692), who wrote in his 1686 *Tabula Chronologica Monarchiae Sinica* that the Buddha had been born 1,036 years before Christ.[34] Jones thus sought to use Buddhist sources (as reported by Europeans) from beyond India to establish the date of the Buddha in India. This Buddha, described in a variety of Hindu texts, especially the Purāṇas, was different from the Buddha that Burnouf and others would later discover in other Sanskrit (and Pāli) works, Buddhist works.

Long after his death, the Buddha had been incorporated into the Hindu pantheon as an incarnation of Viṣṇu. It is said that Lord Viṣṇu appears in the world in each historical age to right a particular wrong, and the specific purpose of his incarnation as the Buddha has been variously described by Hindu authors. Jones explains that Viṣṇu took the form of the Buddha in order to put an end to the sacrifice of cattle, and he quotes a passage from the famous work of the twelfth-century poet Jayadeva, the *Gītāgovinda*. This is Jones's translation of the relevant passage: "Thou blamest (oh, wonderful!) the whole *Veda*, when thou seest, O kind-hearted, the slaughter of cattle prescribed for sacrifice, O *Cesava* [Keśava, an epithet of Kṛṣṇa], assuming the body of *Buddha*: be victorious, O *Heri*, lord of the Universe!"[35] This verse, at least in Jones's rendering, does not offer unambiguous praise of the Buddha, implying that he saw fit to condemn the entire Veda in order to challenge a single element prescribed there: animal sacrifice.

Jones was perhaps moved to translate the passage in this way because of the ambivalence he observed in his brahman informants. He reports, "The *Brahmans* universally speak of the *Bauddhas* [Buddhists] with all the malignity of an intolerant spirit; yet the most orthodox among them consider *Buddha* himself as an incarnation of *Vishnu*."[36] Jones's solution to this apparent contradiction was to propose that there had been two Buddhas. The first according to Jones was named Sacyasinha (Śākyasiṃha, the "Lion of the Śakya Clan," in fact simply one of the epithets of Śākyamuni Buddha); this was the Buddha extolled by Jayadeva as the ninth incarnation of Viṣṇu who prohibited the sacrifice of cattle. At some point after that, "another *Buddha*, one perhaps of his followers in a later age, assuming his name and character, attempted to overset the whole system of the *Brahmans*, and was the cause of that persecution, from which the *Bauddhas* are known to have fled into very distant regions."[37] Thus, the first Buddha, the one worshipped by the Hindus as an incarnation of their god, simply sought to prevent the slaughter of cattle. The second Buddha made a more thoroughgoing attack on Brahmanical authority. As a consequence, his followers were eventually driven from India.

Viṣṇu's purpose in appearing on earth as the Buddha is variously portrayed in the Hindu Purāṇas, but a common narrative is one in which demons gain power through the recitation of the Veda and the practice of asceticism, thus challenging the supremacy of the gods. In order to conquer these demons, Viṣṇu appears as a teacher who condemns the practice of Vedic sacrifice, ignores caste distinction, and denies the existence of a creator deity. The demons become followers of this new teacher, thereby losing their power and being consigned to hell. It was presumably some version of this story that was heard by Jones and his fellow officers of the East India Company, leading to the commonplace that the Buddha opposed the caste system.

The story of the Buddha's incarnation as Viṣṇu is not accepted by any of the Buddhist schools or reported (at least approvingly) in any Buddhist text, and thus would not have been known to the Buddhist monks encountered by Europeans across Asia. Yet two elements of this strange myth (which is hardly an instantiation of the ecumenical spirit, as it has so often been portrayed in neo-Hinduism) quickly attached themselves to the person of the Buddha: his condemnation of animal sacrifice and his rejection of the caste system. It is the latter that concerns us here, but it should be noted in passing that both would play an important role in the European portrayal of the Buddha as a man of the (European) Enlightenment. This characterization of the Buddha's attitude toward caste would persist even after Western scholars gained access to (and the ability to read) Indian Buddhist texts, which presented a rather more nuanced view.

The first detailed discussion of Buddhist attitudes toward caste appeared in 1844 in *Introduction à l'histoire du Buddhisme indien* by the French Sanskritist Eugène Burnouf.[38] The founding of the academic (or in the terms of the day, "scientific") study of Buddhism in the West is often marked by the publication in Paris of this, Burnouf's magnum opus. This 653-page work by the occupant of the chair of Sanskrit at the Collège de France was the first scholarly monograph devoted to Buddhism.[39] As discussed in more detail in chapter 4, in many ways it set the agenda for Buddhist studies for the next century and beyond, in part because Burnouf conclusively demonstrated the Indian origins of Buddhism through an examination of eighty-eight Sanskrit manuscripts that he had received from Brian Hodgson, British Resident at the Court of

Nepal. Prior to that time, scholarship on Buddhism, which appeared for the most part in journals like *Asiatick Researches* and *Journal des Savants*, had been based on works in Chinese, Mongolian, and Burmese, languages of nations where (1) Buddhism still flourished and (2) the caste system did not exist. In India, on the other hand, (1) Buddhism was defunct and (2) the caste system persisted.

Burnouf's extended analysis of caste in Buddhism can only be summarized here. He begins by noting that the Buddha did not reject the fourfold division of society.

> As for the distinction of the castes, in the eyes of Śākyamuni, it was an accident of the existence of humanity here below, an accident that he acknowledged, but that could not stop him. This is why the castes appear, in all the sūtras and all the legends I have read, as an established fact against which Śākya does not make a single political objection. . . . Śākya thus accepted the hierarchy of castes; he even explained it, as the brahmans did, by the theory of sorrows and rewards; and each time that he instructed a man of vile condition, he did not fail to attribute the baseness of his birth to the reprehensible acts this man had committed in a previous life.[40]

For Burnouf, the Buddha's innovation was to declare that the condition of one's birth, the result of karma, was not an impediment to salvation to those reborn in the lower strata of society. "Śākya thus opened to all castes without distinction the path of salvation that before was closed by birth to the greatest number; and he rendered them equal among themselves and before him by conferring on them investiture with the rank of monk."[41] That is, "birth ceases to be a sign of merit as well as a title of exclusion."[42] In Burnouf's view, therefore, the Buddha's target was not the caste system as a whole but only the first caste, the brahmans, the caste of hereditary priests, who by the privilege of birth had sole access to salvific knowledge. The Buddha changed that. "The priesthood ceased to be hereditary, and the monopoly on religious matters left the hands of a privileged caste. The body charged with teaching the law ceased to be perpetuated by birth; it was replaced by an assembly of monks dedicated to celibacy, who are recruited indiscriminately from all classes."[43] The Buddha, therefore, did not seek to transform Indian society by annihilating the classes into which it was divided.

> In order to give substance to his doctrine, Śākyamuni did not need to resort to a principle of equality, in general little understood by the Asiatic peoples. The germ of an immense change was in the constitution of this assembly of monks, coming from all castes who, renouncing the world, had to live in monasteries under the direction of a spiritual leader and under the dominion of a hierarchy based on age and knowledge. People received a most moral instruction from their mouth, and there no longer existed a single human being condemned forever by his birth never to know the truths disseminated by the teaching of the most enlightened of all beings, the perfectly accomplished Buddha.[44]

Burnouf thus seems to perceive a clear distinction between the political and the religious. The *politique* for Burnouf (a term perhaps understood in modern English as "social") encompasses the organization of society into four classes (and many subclasses), resting on two principles: endogamy and hereditary professions.[45] Burnouf provides examples of each without comment, and notes that the Buddha did not seek to disrupt the social order. His aims were instead religious, bringing about a transformation that did not threaten Indian society but which would have been welcomed by the majority of its members.

> The existence of the other castes, on the contrary, was not at all compromised by Buddhism. Founded on a division of work, which birth perpetuated, they could survive under the protection of the Buddhist priesthood, to which they all indiscriminately supplied monks and ascetics. As much as brahmans should feel aversion for the doctrine of Śākya, so much should persons of lower castes welcome it with eagerness and favor; for if this doctrine abased the first, it uplifted the second, and it assured to the poor and the slave in this life what Brahmanism did not even promise in the next, the advantage to see himself, from the religious point of view, as the equal of his master.[46]

Although Burnouf asserts that the principle of equality is "little understood by the Asiatic peoples," he seems to have discerned that principle in the teachings of the Buddha. His own rhetoric is very much that of an anti-Papist French Republican of the first decades of the nineteenth century, proclaiming the rights of all men to enlightenment, and condemning the craven priests who tried to keep it from them. In his diary entry of March 20, 1845, Friedrich Max Müller described his first meeting with his future teacher: "Went to see Burnouf. Spiritual, amiable, thoroughly French. He received me in the most friendly way, talked a great deal, and all he said was valuable, not on ordinary topics but on special. I managed better in French than I expected. 'I am a Brahman, a Buddhist, a Zoroastrian. I hate the Jesuits'—that is the sort of man. I am looking forward to his lectures."[47]

In the history of Buddhist studies in the West, Burnouf's *Introduction* is also important for its many extracts (often quite extended) from Sanskrit Buddhist texts. There is little mention of caste in later Buddhist literature, apart from tantric injunctions to transgress caste boundaries and the often formulaic disputations of the brahmans' claim to supremacy in various treatises on logic.[48] One text in which caste is the central theme is called the *Śārdūlakarṇāvadāna*. Burnouf recounts its framing story in which the Buddha's cousin and personal attendant, Ānanda, encounters a young woman of the *mātaṅga* caste, the lowest of the low, sitting by a well. Ānanda asks her for a drink of water and she replies that she is prevented by her low status from approaching a monk. He replies that he is not concerned about her status, he only asks for water. The young girl, whose name is Prakṛti (meaning "Nature"), immediately falls in love with Ānanda and tells her mother that she wants to be his wife. The girl's mother is a sorceress, and understands that magic will be required to attract the monk to her daughter. As her mother casts a spell, Prakṛti awaits Ānanda in her finest clothes. He is indeed drawn to her, but at the last moment cries out to the Buddha for salvation. The Buddha hears his cry and uses his own magical powers to counteract the mother's sorcery and save the monk from temptation. Prakṛti is not deterred, however, and goes to see the Buddha himself to request permission to be with Ānanda. The Buddha asks her a series of questions: Does she want to go where Ānanda goes? Does she want to eat what he eats? Does she want to wear what he wears? Do her parents want her to be with him? Each of these questions is a double entendre, and the Buddha is in fact conducting the standard interview for admission to the order. And so Prakṛti becomes a nun, but not before the Buddha recites a magical formula (*dhāraṇī*) that purifies the woman of all the evil karma that had caused her to be reborn in her mean condition.[49] If all this sounds like the plot for an opera, it almost was. Wagner read the story in Burnouf's book and reported to King Ludwig II that he was planning an opera called *Die Sieger*, "The Victor," which he never completed.

Burnouf's retelling of the story of Ānanda and Prakṛti, and his astute analysis of the Buddha's attitude toward caste, were read by others as well, including Le Comte de Gobineau (1816–1882), described by Léon Poliakov as "the great herald of biological racism."[50] His most famous work is his admittedly erudite *Essai sur l'inégalité des races humaines*, first published in 1853 in four volumes. His research, he confessed, was "only a means to assuage a hatred of democracy and of the Revolution."[51] In his book, Gobineau sought to demonstrate that the Aryan race, and eventually the entire human species, was doomed because of miscegenation. In a chapter entitled "Buddhism, Its Defeat," he condemns Buddhism for contributing to the demise of the Aryans; although a younger contemporary of Burnouf's, he sees nothing admirable in the Buddha's egalitarian attitude toward caste. The Buddha, he explains, drew followers from the warrior caste, and from the merchants; even some brahmans became his disciples. But "it was chiefly from the base people that he enrolled the majority of his proselytes. At the moment that he threw aside the prescriptions of the Vedas, the separation of the castes no longer existed for him, and he declared that he recognized no other superiority than that of virtue."[52] This for Gobineau was a dark day in Indian history. Indeed, he even attributes the richness of the Buddhist pantheon to this "plunge into the black classes."[53]

But, Gobineau explains, the Buddhists were eventually banished from India, and the brahmans, with their hierarchical caste system, won the day. Buddhism became a nonentity in India, but it went on to China, to Tibet, to Central Asia, that is, to nations without a caste system, where "it can be said that the huge multitudes whose consciences it controls belong to the most vile classes of China and the surrounding countries,"[54] where it purveys "a political and religious doctrine that pretends to be based solely on morality and reason."[55]

In the decades after Gobineau and the rise of race theory, Buddhism would retain its reputation as a religion (or, according to some, a philosophy) "based solely on morality and reason," a religion compatible with science. Such a portrayal, however, would not preclude the injection of the rhetoric of race.

The discourse on race and caste in Buddhism during the nineteenth century was not confined to France. In the last half of that century, a movement that has retrospectively been dubbed "Buddhist modernism" began. One of the several family resemblances of its various manifestations was the emergence of Asian elites who adopted the European representation of Buddhism, at least the more romantic aspects of it, and then put it to use in defending the dharma against both colonialism and missionary Christianity. The most prolific and visible of the Buddhist modernists was named Dharmapāla.

Anagārika Dharmapāla (1864–1933), already encountered in the introduction, was born Don David Hewaviratne in Sri Lanka, at that time the British colony of Ceylon. He was raised in the English-speaking upper class of Colombo and was educated in Catholic and Anglican mission schools, where he is said to have memorized large portions of the Bible, although his family was Buddhist. In 1880 he met Madame Helena Petrovna Blavatksy and Colonel Henry Steel Olcott, founders of the Theosophical Society, during their first visit to Ceylon. In 1884 he was initiated into the Theosophical Society by Colonel Olcott, and later accompanied Madame Blavatsky to the society headquarters in Adyar, India. She encouraged him to study Pāli, the language of the Theravāda Buddhist scriptures, after his return to Ceylon.

In 1881, soon after meeting Blavatsky and Olcott, he took the name Dharmapāla, "Protector of the Dharma." Prior to that time in Sri Lanka, the leadership in Buddhism had been provided by monks and kings. Dharmapāla sought to establish a new role for Buddhist laypeople, proclaiming the revival of the order of the *anagārika*, or wanderer. In fact, he created a category of which he was the sole member: that of a layperson who studied texts and meditated, as monks did, but remained socially active in the world, as laypeople did.

Dharmapāla's name requires brief comment. The canonical language of the Buddhism of Sri Lanka, the land of Dharmapāla's birth, is Pāli, a vulgate of Sanskrit that is regarded by the monks of the Theravāda tradition of Sri Lanka and Southeast Asia as the original language of Buddhism, and of the Buddha himself. In the late nineteenth century, this was the considered scholarly view in Europe as well (since abandoned). Yet the constituents of the name chosen by the young Don David are not Pāli. The first constituent of his new name *anagārika* is a Sanskrit term meaning "homeless," and it appears more commonly in Hindu than Buddhist contexts, where the preferred form is *anagāriya*. The second constituent, *dharmapāla*, "protector of the dharma," is also a Sanskrit term. The Pāli term would be *dhammapāla*. Unlike the rarely used *Anagārika*, *Dhammapāla* is a well-known name in the Theravāda tradition. The Sanskrit version, Dharmapāla, is also a common surname in Sinhala; Sanskrit names were sometimes used by Sinhalese Buddhists, in part because they carried a certain cultural cachet that Pāli lacked. Don David selected two Sanskrit terms for his title, leading one to conjecture that these terms came to him not from the Pāli tradition that he would exalt throughout his career (although apparently without learning to read it very well) but from Sanskrit. That he took this name during his early days of devotion to Blavatsky (who generously seasoned her tomes with Sanskrit terms) suggests the possibility that it may have been chosen under her influence.

In 1886 Edwin Arnold (1832–1904), author of the best-selling biography of the Buddha, *The Light of Asia*, visited Sri Lanka. He had recently been to Bodh Gayā, the site of the Buddha's enlightenment, which for centuries had been under the control of Hindu priests. Shocked by the sad state of "the Buddhist Jerusalem," he recounted his experience in an article in the *Daily Telegraph*; when he had asked a Hindu priest whether he might pick a few leaves from the Bodhi tree, the priest replied, "Pluck as many as ever you like, sahib, it is nought to us."[56] At a meeting with a group of Sri Lankan Buddhists (which included Dharmapala), Arnold called for the Buddhist sites of Bodh Gayā and Sarnath (where the Buddha had preached his first sermon) to be placed under Buddhist control.

In 1889 Dharmapāla accompanied Colonel Olcott on a lecture tour to Japan. He was finally able to visit Bodh Gayā in 1891, where he was shocked to see the state of decay of the most important Buddhist pilgrimage site of India. That same year, he helped to found the Maha Bodhi Society, which called on Buddhists from around the world to work for the restoration of the site of the Buddha's enlightenment to Buddhist control.

In 1893 Dharmapāla attended the World's Parliament of Religions, held in conjunction with the Columbian Exposition in Chicago. Although one of several Buddhist speakers, his excellent English and Anglican education made him an eloquent spokesman for the dharma, demonstrating both its affinities with and superiority to Christianity in a speech entitled "The World's Debt to Buddhism." A letter to the editor of the *St. Louis Observer* on September 21, 1893, offered this description:

"With black curly locks thrown from the broad brow, his clean, clear eye fixed upon the audience, his long brown fingers emphasizing the utterances of his vibrant voice, he looked the very image of propagandist, and one trembled to know that such a figure stood at the head of the movement to consolidate all the disciples of Buddha and spread 'the light of Asia' throughout the civilized world."[57]

Dharmapāla was a prolific essayist and pamphleteer, writing on a wide range of subjects; his collected speeches, essays, and letters fill almost nine hundred pages, and these include only his writings in English; he also published extensively in Sinhala. He was a strong critic of the British administration of the colony of Ceylon, leading to his being labeled as a "notorious seditionist" (and a homosexual) by British colonial officers and interned in Calcutta from 1915 to 1920.[58] Undaunted, Dharmapāla would come to see his homeland of Sri Lanka as a site of a concentrated purity, refined from three sources: language, race, and religion, each of which he would describe with the word *Aryan*.

By the late nineteenth century, Aryan discourse had made its way to Sri Lanka. In his 1855 *The Languages of the Seat of War in the East with a Survey of the Three Families of Language, Semitic, Arian, and Turanian*, F. Max Müller had observed that Sanskrit "has ceased to live, and though it exists still like a mummy dressed in its ceremonial robes, its vital powers are gone. Sanskrit now lives only in its offspring; the numerous spoken dialects of India, Hindustani, Mahratti, Bengali, Guzerati, Singhalese, &c., all preserving, in the system of their grammar, the living traces of their common parent."[59] In a paper read before the Ceylon Branch of the Royal Asiatic Society on October 31, 1863, the Sinhalese Protestant scholar James de Alwis (1823–1878), citing this passage from Müller, asserted that Sinhala belonged to the northern and Aryan class of languages, rather than the southern and Dravidian.[60] In the same essay, he drew the connection between language and race.

> The colour as well as the features of the inhabitants of the Dekkan are certainly distinguishable from those of the Sinhalese even by a casual observer. An utter stranger to the various races cannot be three weeks in this Island before he perceives the striking difference between the manners and habits of the Sinhalese on the one hand, and those of the different other races on the other. European Teachers have frequently observed the facility with which the Sinhalese pronounce European tongues, pre-

senting in this respect a quality distinguishable from every race of South-Indian people.[61]

The British legal historian Henry Sumner Maine (1822–1888), who in his famous 1861 treatise, *Ancient Law*, had compared Hindu and Roman law codes, returned from six years in India as Vice Chancellor of the University of Calcutta to declare in his Rede Lecture of 1875, "the new theory of Language had unquestionably produced a new theory of Race."[62] Maine's first major work after his return to England, *Village-Communities in the East and West* (1871), was employed by British officials in Ceylon in drafting the Ceylon Village Communities Ordinance.[63] The leap from language to race was made explicitly by Mahadeva Moreshwar Kunte (1835–1888) in his 1879 *Lecture on Ceylon*, in which he declares that there are "only two races in Ceylon—Aryans and Tamilians—the former being divided into descendants of Indian and Western Aryans."[64]

Thus, by the last half of the nineteenth century, the inhabitants of the colony of Ceylon, long divided into the Sinhalese and the Tamils, had been identified as Aryans and Dravidians. The role of the Aryan race in the history of world civilization had also been described. Writing on leadership in his influential 1877 *Ancient Society; or, Researches in the Lines of Human Progress from Savagery, through Barbarism to Civilization*, Lewis Henry Morgan (1818–1881) states that "from the Middle Period of barbarism, however, the Aryan and Semitic families seem fairly to represent the central threads of this progress, which in the period of civilization has been gradually assumed by the Aryan family alone."[65]

Victorian race science was easily accommodated to Sinhalese myth. The Sinhalese traced their origin to the myth of Vijaya, an Aryan king of northern India who sailed from the city of Sinhapura in Bengal to settle on the island of Laṅka. As Dharmapāla wrote in 1902,

> Two thousand four hundred and forty-six years ago a colony of Aryans from the city of Sinhapura, in Bengal, leaving their Indian home, sailed in a vessel in search of fresh pastures, and they discovered the island which they named Tambapanni, on account of its copper coloured soil. The leader of the band was an Aryan prince by the name of Wijaya, and he fought with the aboriginal tribes and got possession of the land. The descendants of the Aryan colonists were called Sinhala, after their city, Sinhapura, which was founded by Sinhabahu, the lion-armed king. The

> lion-armed descendents are the present Sinhalese, whose ancestors had never been conquered, and in whose veins no savage blood is found. Ethnologically, the Sinhalese are a unique race, inasmuch as they can boast that they have no slave blood in them, and never were conquered by either pagan Tamils or European vandals who for three centuries devastated the land, destroyed ancient temples, burnt valuable libraries, and nearly annihilated the historic race.[66]

Vijaya was said to have landed on the island on the day that the Buddha passed into nirvāṇa. Before doing so, however, the Buddha asked Śakra, the king of gods, to protect Vijaya's sojourn in Sri Lanka, where the dharma would flourish for five thousand years.[67]

In fulfillment of this prophecy, the third source of Sinhalese purity arrived on the island 236 years after the Buddha's passage into nirvāṇa, when the pious monk Mahinda, son of the emperor Aśoka, came from India and established the saṅgha on the island, the island that the Buddha was said to have visited on three occasions, the island that would come to be known as *dhammadīpa*, the Island of the Dharma.

For Dharmapāla, then, Sri Lanka was triply Aryan, ennobled by its language, its race, and its religion. It was as if the Indian Subcontinent were a funnel, with the Aryan language, the Aryan blood, and the Buddhist dharma of the north trickling south to be concentrated and preserved in their purest form in the island at the funnel's tip. The subsequent inhabitants of the island, the Tamil Hindus and the Muslims (or "Moors," as he called them), were not true Sinhalese because they were not Aryan in language, in race, in religion.[68] He wrote in 1915, "What the German is to the Britisher that the Muhammedan is to the Sinhalese. He is an alien to the Sinhalese by religion, race and language. He traces his origin to Arabia, whilst the Sinhalese traces his origin to India and to Aryan sources."[69] Elsewhere, he informs the "young men of Ceylon" that "by religion, by race, by traditions, by our literature we are allied to the Aryan races of the Gangetic Valley."[70]

Dharmapāla's discourse appears to be the familiar product of the nineteenth century's transition from philology to race science, when the branches of the family trees of languages turned into bloodlines. What is striking here is the source: the language of race and racial purity is spoken by an Asian Buddhist, and in the name of the Buddha. And that language is spoken with an increasing level of vituperation, not merely against the traditional (and in many senses deserving) target of the Christian missionary, but against Christianity itself, its founder, and ultimately all that was not Aryan, a term that for him came to be synonymous with Buddhism.

Dharmapāla championed the general superiority of Indian civilization over that of Europe. Indian civilization is older, and more refined, than that of Europe, which, without a civilization of its own, was susceptible to the primitive beliefs of desert tribesmen.

> Remember India is a continent, not like Palestine or Arabia, peopled by wild, roving Semitic Bedouins, children of the desert, and that it is a vast country peopled by highly spiritualized races, with a civilization going back to thousands and thousands of years, and the cradle land of religions and philosophies. In a country where religious inquiry is man's birthright, dogmatism has no place. India never knew in its long record of history to persecute people for their religious opinions. The persecuting spirit of religious tyranny began with the Semitic Jehovahism, and later ruthlessly followed by the founder of Islam. The Semitic spirit was implanted in the Latin and Teuton heart after the introduction of the Semitic doctrine of Palestine into Europe. Never having had a religion with a history and theology among the European races, it was a [*sic*] quite easy for the promulgators of the Semitic faith to impress on the European mind the terribleness of the Jealous Jah of Mt. Horeb. Europe succumbed, and its future was made a blank by means of terrifying dogmatism ending with hell fire and brimstone to eternity.[71]

If Buddhism has an analog in the West for Dharmapāla, it is the ancient, pre-Christian civilizations of Greece and Rome, before the foreign slaves of the imperial Rome converted to Christianity: "The slaves accepted the teachings of Jesus since they suited the slave temperament."[72] Prior to this, however, "The ancient Greeks thought like the ancient Aryans of India, the gods they worshipped were not of the semitic type. . . . The draped figures of the Greek poets and philosophers were exact representations of the statues of the ancient Aryan Bhikkhus."[73] "Roman and Grecian civilization originally was Oriental. The religions they professed were not Semitic."[74]

In his criticism of colonialism, Dharmapāla traced the violence and rapaciousness of the European powers back to their religion. He wrote in 1926, "With the exception of Buddhism all religions have been destructive. In India Brahmanism partially destroyed Buddhism, and the remnant of Buddhists that existed was destroyed by the Muslims seven centuries ago. . . . Today England, the United States, Italy, Belgium, France, Germany and other Christian countries are sending shiploads of missionaries to China, Ceylon, India, Japan, Burma to convert both Buddhists and Hindus, backed by the capitalists and gunboats."[75] In discussing the case of his own country of Sri Lanka in 1923, he found the religion preached by these missionaries, and indeed all "Western" religions, to be illegitimate.

> The Ceylon Buddhists that have succeeded in maintaining Buddhism for a period of 2200 years in the Island are now confronted with the sensualistic creeds of Mecca and Palestine. Judaism is a down-right plagiarism. It has robbed from Babylonian religions, Assyrian religions, Egyptian religions, Zoroastrianism, the doctrines that were current in the Euphrates valley and in Persia. Its bastard offshoot has borrowed a large stock of ethics from Buddhism. The ceremonialism of the Byzantine Christian Church was copied from the Buddhism of Turkestan and Turfan.[76]

Dharmapāla's essays, then, are marked by images of the authentic and the inauthentic, the legitimate and illegitimate, the pure and the polluted, with the authentic, the legitimate, and the pure described as Aryan. He refers to the Buddha as the great "Aryan Saviour" who taught "the noble Aryan people of India."[77] In doing so, he is clearly aware of the ways in which *Aryan* has taken on a racial connotation in Europe, one that, in an apparently more benign form, linked Britain and India. Max Müller had famously written in 1855 that no authority could "convince the English soldier that the same blood was running in his veins, as in the veins of the dark Bengalese. And yet there is not an English jury now a-days, which, after examining the hoary documents of language, would reject the claim of a common descent and a legitimate relationship between Hindu, Greek, and Teuton."[78] Dharmapāla was not interested in such brotherhood, however, and stated as much in 1924:

> The British people today take pride in calling themselves Aryans. There is a spiritualized Aryanism and an anthropological Aryanism. The Brahmans by enunciating a system of Griha Sutras called those people only

Aryans who lived in the territory known as Bharatvarsha. Those who did not conform to the sacred laws were treated as Mlechhas.

Buddhism is a spiritualized Aryanism. The ethics of the Bible are opposed to the sublime principles of the Aryan Doctrine promulgated by the Aryan Teacher. We condemn Christianity as a system utterly unsuited to the gentle spirit of the Aryan race.[79]

Dharmapāla here alludes to the two apparent meanings of *Aryan*. The British are not Aryan from the Hindu perspective because, regardless of their colonial occupation of India, they are not native to the soil of the Bharatavarṣa, the ancient Sanskrit name for India. Furthermore, the British do not follow the ancient law codes of India. They are thus *mleccha*, barbarians. They are also not Aryan in the Buddhist sense because their religion is contrary to the ethical and thus ennobling teachings of the Buddha.

Dharmapāla appears to confuse his metaphors of the material and spiritual when he refers to "the gentle spirit of the Aryan race." Or perhaps not. The most eloquent apostle of Buddhism of his age (or at least the most eloquent English-speaking apostle), Dharmapāla was concerned to demonstrate that Buddhism is a world religion, equal, indeed superior, to all others. The Aryan, literally "superior," nature of Buddhism could not then be simply a matter of language or bloodline. It had instead to be a spirit in which all Buddhists of Asia partook. "The Bhikkhus wearing the yellow robe of purity and love went to distant lands to spread the ethics of Aryan culture. They Aryanized the unaryan races."[80] Buddhism may indeed be found in its purest form among Aryan peoples of Sri Lanka and in the Aryan language of Pāli. But Buddhism, for Dharmapāla, also had to be universal, and hence "a spiritualized Aryanism." It had long been a tenet of the Theravāda traditions of Sri Lanka and Southeast Asia that so much time had passed since the passing of the Buddha that it was no longer possible to become an arhat, the ultimate *āryan*, who would enter nirvāṇa at death. The *āryan*, then, from the perspective of Buddhist doctrine, was the rarest of saints. Dharmapāla sought to restore the status of the *āryan* and expand it to encompass all Buddhists, a Buddhist nation unbounded by national borders, superior to those of the other great religions.

He thus did not hesitate to contrast Buddhism with the two other "universal religions" in the bald language of race: Christianity and Islam "belong to the Semitic family, while Buddhism belongs to the Aryan family."[81] And from these origins, the two traditions diverged. "The two Semitic religions that had their origin in Arabia and Palestine are responsible for the retardation of progress of the larger Humanity of Asia. Islam is responsible for the destruction of the Aryan civilization of India. All that was beautiful in aesthetic architecture, built by the devotees of Aryan spirituality, went down with a crash, under the sledgehammer attack of semitic monotheism."[82] Or as he put it more succinctly in 1932, "Within the last thousand years, what cruelties have the followers of these two religions not committed in the name of a Semitic deity."[83] Europe, he explains, "received its religion from the Asiatic Jews."[84]

Mutually flattering comparisons of Christianity and Buddhism have been a standard element of Buddhist modernism since the nineteenth century, especially those that compare Jesus and the Buddha. Edwin Arnold followed his biographical poem about the Buddha, *The Light of Asia*, in 1879, with a poem about Jesus, *The Light of the World*, in 1891. Works comparing Jesus and the Buddha continued to be published over the next century and to the present day. Although Dharmapāla at times writes approvingly of Jesus and his teachings, elsewhere he is scathingly dismissive, calling Jesus "half Jew and half Hittite," who taught "a hotch potch mixture of Judaism, Brahmanism, and Buddhism."[85] Jesus was "camouflaged as the prince of peace, whilst his actions show him to be a personality with an irritable temper. His very disciples forsook him at the critical moment when he prayed for help. He died praying to his god confessing his ignominious failure."[86] "He was an exorcist. He taught no philosophy. He followed the profession of an African rain-doctor"[87] and "showed that he was devoid of compassion by his cruel behaviour in sending 2000 hogs to drown in the sea. . . . He was rude towards his mother, not once, but thrice. Perhaps he was angry with his mother because she could not tell him whose son he was."[88] Jesus shows "his Jewish nature" by recognizing usury in the parable of the talents. He condemns to hell those who "failed to believe he is the son of the Arabian god, Jehovah."[89]

Jesus and the Buddha were thus utterly different. The Buddha, whom Dharmapāla refers to as "the great Aryan Saviour," was "the great Conqueror who spent forty five years of His perfect manhood in the moral regeneration of the greater part of advanced Humanity."[90] Jesus, however, only "had eleven disciples, men of low intelligence, and his congregation was the riff-raff of Galilee—the backwash of the barren parts of Asia. The life of Jesus was an absolute failure. He was a victim of megalomania and at times suffered from paranoia," rendering his teachings so different from "the virile, vigorous manly ethics of the Tathagato."[91]

In the late nineteenth and early twentieth centuries, there was a high level of anxiety in Europe about the Semitic origins of Christianity, and Buddhism was called to the rescue by some. Yet in those cases, Jesus is usually somehow spared the critique, often with one account or another of where he spent those lost years. Rarely does one find Jesus characterized as a glutton and a drunkard, as we find in Dharmapāla.

Dharmapāla was of course one of the early proponents of Buddhism as scientific, and his claims in this domain were also cast in the rhetoric of race. He sometimes referred to Buddhism as "Aryan psychology," and repeatedly contrasted the Buddha's rational spirit of inquiry and his rejection of priestcraft long before Christian priests persecuted Bruno and Galileo. In contrast to the benighted views of Christian theology with "its unscientific doctrines of creator, hell, soul, and atonement,"[92] the Buddha, "a scientist full of compassion for all,"[93] had set forth the principles of a universal religion twenty-five hundred years ago, and "the panacea needed to cure the muddleheaded is still to be found in the laboratory of the great Aryan Teacher."[94] Indeed, "the discoveries of modern science are a help to understand the sublime Dhamma."[95]

It is clear from the foregoing that Dharmapāla's conception of science also included the ignoble science of race, which he used repeatedly to bludgeon the powers that opposed him. Like the larger Buddhism and Science discourse, Dharmapāla's embrace of the science of race was a product of colonialism. Indeed, it could be argued that the Sanskrit term *āryan*, a term that he invoked so often, was not, at least in the sense that he used it, a Buddhist term at all. It was instead a colonial commodity, produced in India millennia ago, refined in Europe in the nineteenth century yet left in Sanskrit untranslated, then exported back to India. There it was first used to name the natives, as the German Romantics and early Indologists like Burnouf extolled the spiritual heritage of India as the inheritance of Europe. As South Asia came increasingly under British rule in the nineteenth century, *āryan* became a precious artifact of a classical past, long lost in a process of invasion, miscegenation, and decline, better preserved in Europe than in India. Dharmapāla was one of the consumers of this commodity in the British colony of Ceylon, just off the coast of India, where he made it his own, turning it into a weapon against the European oppressors and their religion. In order to do so, however, he resorted to a rhetoric that strikes our ears as distinctly non-Buddhist, even un-Buddhist. The final question, perhaps, is whether he was giving voice, however ignobly, to something that had been there all along, something that, like the low-caste woman named Nature, must be purified by the Buddha's magic spell in order to be admitted to his noble culture as it is represented today.

• • •

Perhaps the most renowned of all Buddhist philosophers is Nāgārjuna. Also known as Ārya Nāgārjuna, his most famous work is the *Madhyamakaśāstra* (Treatise on the Middle Way), which begins with the famous line, "I bow down to the perfect Buddha, the best of teachers, who taught that what is dependently arisen is without cessation, without production, without annihilation, without permanence, without coming, without going, without difference, without sameness, pacified of elaboration, at peace." Of the many commentaries on this text, one of the most important is that of the sixth-century author Bhāvaviveka. This is his exegesis of the passage:

> The compassionate and excellent Bhagavan, endowed with the intelligence that creates joy, took upon himself "the immeasurable burden," the gathering of merit and wisdom for measureless myriads of aeons. In doing so, he created benefit for others and was endowed with the intention that was not disheartened even by the suffering of losing his own life. Seeing the world wandering aimlessly in the thick net of elaborations like production, cessation, annihilation, permanence, coming, going, difference, and sameness, he churned the ocean of objects of knowledge with his wisdom and, without depending on another and without conceptions, obtained the ambrosia of the reality of all phenomena, well free of the nets of elaboration. By way of the supreme vehicle, he set forth to transmigrators precious and excellent dependent origination, using expressions like "production" and "non-production," relying on conventional and ultimate truths, unshared by the *tīrthika*s, *śrāvaka*s, and *pratyekabuddha*s, without overstepping [appropriate] birth, age, lineage, place, and time.[96]

Bhāvaviveka's commentary on Nāgārjuna's sentence does not warrant particular comment; it provides a standard Mahāyāna reading of the Buddha's compassionate teaching, alluding in the last phrase to his storied "skillful methods" (*upāyakauśalya*) in which he teaches in a way that is appropriate to the interests and capacities of his audience.

But the particular list of "birth, age, lineage, place, and time" is somewhat unusual. Commentary sometimes, therefore, requires commentary. For this we turn to one of the most detailed, and lengthy (at almost two thousand folio sides), commentaries in Indian Buddhist literature, the subcommentary on Bhāvaviveka's commentary by Avalokitavrata, who presumably lived in the seventh century. Here is how he explains the terms:

> Regarding this precious and excellent dependent origination, [it is taught] without overstepping birth, age, lineage, place, and time. It does not diminish or overburden those. Regarding "without overstepping birth," it is to be taught to those who have been born as gods or humans. It is not to be taught to those who have been born as hell beings, animals, or ghosts. In that way, it does not transgress birth. Regarding "without overstepping age," it is be to taught to one in the prime of life. It is not to be taught to one who is very young or one who is very old. In that way, it does not transgress age. Regarding "without overstepping lineage," it is to be taught to brahmans or kṣatriyas; it is not to be taught to vaiśyas or śūdras. As a text of the outsiders says: "One should not impart knowledge to a śūdra, nor should one give him leftover food, nor the remains of a sacrificial offering. One should not teach him the dharma, nor teach him religious practices." It is without such transgressions. Regarding "without overstepping place," it is to be taught to one born in a central country and in a town or a monastery; it is not to be taught to one born in a borderland or in a charnel ground or at a crossroads. In that way, it does not transgress place. Regarding "without overstepping time," it is to be taught in the morning or the night; it is not to be taught at twilight. In that way, it does not transgress time.[97]

This passage is translated from the Tibetan; it is lost in the original Sanskrit. The term translated as "lineage" (*rigs*) is probably *varṇa* in the Sanskrit. The passage from "a text of the outsiders" (that is, non-Buddhists) is the eightieth verse of the fourth chapter of the *Laws of*

Manu, the same passage cited at the beginning of this chapter. This raises the obvious question of how consistent the Buddhist critique of caste was in India, when we find an Indian Buddhist scholastic, a proponent of the Mahāyāna (rendered by some as "Universal Vehicle"), citing, with approval, the statement from the *Laws of Manu* that the dharma, sometimes rendered as "truth," should not be taught to a śūdra.

TWO TIBETANS

The discourse of Buddhism and Science seems to have been largely, if not entirely, absent in Tibet in the last decades of the nineteenth century when the relation of Buddhism and Science captured the attention of Buddhist elites elsewhere in the Buddhist world. There are the accounts of Tibetans who journeyed to Nepal or India and reported on the wonders of the modern world they saw there.[1] There is the report of the Bengali scholar Sarat Chandra Das who entered Tibet as a British agent and was befriended by the Seng chen Tulku (Blo bzang dpal ldan chos 'phel), the abbot of Dongtse ('Brong rtse) Monastery and chief minister (*skyabs dbyings*) of the Ṣixth Panchen Lama. Das reports that this prominent monk showed great interest in the camera and the magic lantern, and may have written about them in Tibetan. As a consequence of his contact with Das, the Seng chen Tulku was arrested, imprisoned, publicly flogged, and then brutally drowned in 1887. Upon his death, his line of incarnation was terminated by the government of the Thirteenth Dalai Lama, who was then in his minority. But apart from these, there are few references to science in Tibetan literature of the nineteenth century.

Two Tibetans, indeed perhaps the two most famous Tibetans of the twentieth century, wrote about Buddhism and Science some decades after the discourse had begun elsewhere in Asia: in Sri Lanka, China, and Japan. It is noteworthy that neither of these Tibetans wrote about Buddhism and Science when they lived in their native land; both did so while in India. The first, Gendun Chopel, wrote at some length on the topic in the late 1930s, some fifty years after the execution of Seng chen Tulku, although his words would not be published for another fifty years. The second, the Fourteenth Dalai Lama, became interested in science during his trips to India and China before the fall of Tibet in 1959; his interest has grown in exile. An examination of their insights into the question of Buddhism and Science provides a powerful lens through which a host of important questions come into focus.

Gendun Chopel (Dge 'dun chos 'phel) was born in Amdo, the northeastern region of Tibet, in 1903. His father was a Nying ma lama, and after his birth Gendun Chopel was himself identified as the incarnation (*sprul sku*) of a prominent Nying ma lama. His intellectual gifts led him at age fourteen to enter the local Geluk monastery with five hundred monks, called Diza (Rdi tsha), and he would henceforth be known to many as *rdi tsha skam po*, "Diza Slim." In 1920 he moved to the great regional monastery of Labrang (Bla brang bkra shis 'khyil), where he quickly gained notoriety as a skilled debater. He was eventually forced to leave the monastery under somewhat mysterious, but apparently unfavorable, circumstances. In 1927 he set off on the four-month trek to the capital of Tibet, Lhasa, where he enrolled at Drepung ('Bras spung) Monastery, renowned for having 7,700 monks, but with a monastic population of between 10,000 and 12,000 at that time. He seems to have completed the scholastic curriculum but did not stand for the geshe examination, which would have conferred the highest academic rank of the Geluk academy. In 1934 he accompanied the Indian pundit Rahul Sankrityayan (1893–1963) in his search for Sanskrit manuscripts in the monasteries of southern Tibet. He ended up accompanying Pundit Rahul to Nepal and then on to India, where he would spend the next twelve years.

Gendun Chopel was extremely active during this period. He traveled extensively, often alone, throughout South Asia, from Kalimpong in the north to Sri Lanka in the south. He studied Sanskrit, Pāli, and English, gaining various degrees of facility in each. He translated the famous Sanskrit drama *Śakuntalā* and several chapters of the *Bhagavad Gītā* into Tibetan and is said to have translated Dharmakīrti's great work on Buddhist logic, the *Pramāṇavarttika*, from Sanskrit into English, in collaboration with an unidentified Christian nun; this translation is not extant. He assisted the Russian Tibetologist George Roerich in the translation of the important and largely unreadable fifteenth-century history of Tibetan Buddhism, the *Blue Annals* (*Deb ther sngon po*). He was given access to several Dunhuang manuscripts on the Tibetan dynastic period by the French Tibetologist Jacques Bacot, and he read translations of Tang historical records. He eventually used these Tibetan and Chinese works as the basis for his unfinished political history of Tibet, the *White Annals* (*Deb ther dkar po*). He visited and studied most of the important Buddhist archaeological sites in India, writing a pilgrimage guide that is still used. And he studied Sanskrit erotica and frequented Calcutta brothels, producing his famous sex manual, the *Treatise on Passion* (*'Dod pa'i bstan bcos*).

Gendun Chopel spent the last two years of his travel abroad, from 1944 to 1946, in northern India and Sikkim, where he became involved in discussions with a small group of Tibetans who would become the ill-fated Tibet Improvement Party. He returned to Tibet in early 1946. In Lhasa in late summer, the government placed him under arrest, informing him only that charges had been brought against him for distributing counterfeit currency. He maintained his innocence throughout his interrogation but was incarcerated, eventually in the prison at Zhol, at the foot of the Potala.

He was released in 1949, just a year before the Chinese invasion. By most accounts, he emerged from prison a broken man and became increasingly addicted to alcohol. His writings had been confiscated and he showed no interest in reviving his many projects, although he may or may not have dictated what would be his most controversial work, the *Adornment for Nāgārjuna's Thought* (*Klu sgrub dgongs rgyan*), to a disciple.[2] He died of undetermined causes in October 1951, at the age of forty-eight.

During his years in South Asia, Gendun Chopel contributed poems and essays to *Melong* (the *Tibet Mirror*), the only Tibetan-language newspaper, published in Kalimpong in northern India by the Tibetan Christian Babu Tharchin (1890–1976). In the June 28, 1938, issue, he published the essay entitled "The World Is Round or Spherical" under the byline "Honest Dharma." That essay was discussed in chapter 1.

Gendun Chopel's most sustained discussion of science occurs in his longest work, *The Golden Chronicle, the Story of a Cosmopolitan's Pilgrimage*

(*Rgyal khams rig pas bskor ba'i gtam rgyud gser gyi thang ma*), which he regarded as his most important contribution to Tibetan knowledge. Composed and illustrated between 1934 and 1941, this work is often described as Gendun Chopel's "travel journals." In fact, it is much more than that, representing his encounter and conversation with traditional Indian culture as well as with modernity, as they appeared to him in colonial South Asia. His discussion of science is found in the concluding chapter of this long book (606 pages in the original Tibetan), completed in Sri Lanka in 1940 and 1941. Simply entitled "Conclusion" (*mjug rtsom*), it consists of a series of reflections on the state of affairs in the India of his day. It begins with a powerful critique of European colonialism and ends with Gendun Chopel's rather wishful claims that Buddhism can still be found in India, if one only knows where to find it. The discussion of science is not marked as a special section of the chapter, falling as it does between a description of Madame Blavatsky and a description of Mahatma Gandhi. At the end of the passage on Blavatsky, he writes:

> It fascinates all the westerners because she explains her religion by stitching it together with the views of modern science. In particular, in the past, there were foreigners who did not believe in the supernatural. Not only did she demonstrate magic to them but she applied scientific principles to such things as transforming matter through magical powers. That mode of explanation seemed to impress everyone. However, if it were explained to us [Tibetans] who are not familiar with the assertions of science, it would only confuse us.[3]

His allusion to science seems to prompt him to take up the topic in more detail. It is a lengthy passage, certainly the most detailed engagement with the question of the relation of Buddhism and Science by a Tibetan (and in Tibetan) before the Dalai Lama's statements in the 1990s. Because of this significance, I will provide a translation of the entire passage before offering a detailed examination of Gendun Chopel's arguments.

> Now I offer a sincere discussion for those honest and far-sighted friends who are members of my religion. The system of the new reasoning "science" is spreading and increasing in all directions. In the great countries, after scattered disagreements among many people, both intelligent and

stupid, saying, "It is not true," they all have become exhausted and must fall silent. In the end, even the Indian brahmans, who care more about a literal interpretation of [their] scriptures than their lives, have had to powerlessly accept it.

These assertions of the new reasoning are not established through one person arguing with another. For example, a spyglass constructed by new machines sees something thousands of miles away as if it were in the palm of one's hand, and similarly, a glass that sees what is close by makes even the smallest atoms appear the size of a mountain; one can analyze the myriad parts, actually seeing everything. Therefore, apart from closing their eyes, they [the opponents of science] had nothing else to try. At first, even those who adhered to the Christian religion in the foreign lands joined forces with the king, casting out the proponents of the new reasoning, using whatever methods to stop them, imprisoning them and burning them alive. In the end, just as one cannot hide the sunshine with one's hands, so also the parts of their religion that were unacceptable within the new system were defeated and they admitted that they were utterly false. As the glorious Dharmakīrti said [at *Pramāṇavartikka* 1.221], "Those who are mistaken about the truth cannot be changed, no matter how one tries, because their minds are prejudiced." The rejection of reason is the most despicable act.

Even so, when we [Tibetans] hear the mere mention of the new system, we look wide-eyed and say, "Oh! He is a heretic!" There is the danger that when we come eventually to believe impulsively in the new reasoning, we will lose all faith in the Buddha, like some Mongolian of the Urga region [i.e., Communists], and become non-Buddhists. Therefore, whether one either stubbornly says "No!" to the new reasoning or believes in it and utterly rejects the teaching of Buddhism, both are prejudices; because it is simply recalcitrance, this will not take you far.

No matter what aspect is set forth in this religion taught by our teacher [the Buddha], whether it be the nature of reality, how to progress on the path, or the good qualities of the fruition, there is absolutely no need to feel ashamed in the face of the system of science. Furthermore, any essential point [in Buddhism] can serve as a foundation for science. Among the foreigners, many scholars of science have acquired a faith in the Buddha, becoming Buddhists, and have even become monks.

One person has told me: "First, I followed the system of the ancient religion of Jesus. Later, I learned science well and a new understanding was born. Then I thought that all the religions in this world are just assertions rooted in a lie, requiring that one rely only on the letter. One day, having seen a stanza of the dharma translated into a foreign language, I thought, 'Oh! The only one who follows the path of reason is the Buddha. Not only did he climb the ladder of science, but having left that [ladder] behind, he traveled even further beyond,' and conviction was born."

Recently a famous renunciant named Trailokajñāna living in Sri Lanka said that in the future the religion of the Buddha will be called the religion of science, that is, a religion of reason, and other religions will be called religions of faith. Another Buddhist paṇḍita says, "Having mastered scientific reasoning, I came to believe in the Buddha. The religion of my teacher works hand in hand with scientific reasoning; when one side tires, the other is able to leap over [to assist]. If other religions join hands with science, they collapse, either immediately or after a few steps."

For example, the followers of the new reasoning say, "In the second moment immediately after any object comes into existence, it ceases or dissolves. These collections of disparate things disperse like lightning." Consequently, the first moment of a pot does not persist to the second moment, and even the perception of a shape does not exist objectively apart from the power of mind or of human language. Moreover, when examined as above, even colors are merely the ways a wave of the most subtle particles moves. For example, regarding waves of light, there is no difference of color whatsoever to be seen in the particles that are the basis for that color; it is simply that eight hundred wavelengths in the blink of an eye appear as red and four hundred appear as yellow. Furthermore, they have invented another apparatus for seeing things that move too quickly to be seen, like drops of falling water. Something that lasts for one blink of an eye can be easily viewed over the duration of six blinks of an eye. More than ten years have passed since they made a viewing apparatus that is not obstructed [in seeing] things behind a wall or inside of a body. All of this is certain. They have also made a machine by which what is said in India can be heard in China in the following moment. Because they are able to show in China a film of something that exists in India, all people can be convinced. The final proof that all things run on waves of electricity is seeing it with one's own eyes.

Many great scholars of science made limitless praises of the Buddha, saying that two thousand years ago, when there were no such machines, the Buddha explained that all compounded things are destroyed in each moment and he taught that things do not remain even for a brief instant, and subsequently we have actually seen this using machines. The statement by Dharmakīrti that "continuity and collection do not exist ultimately" can be understood in various ways, but in the end one can put one's finger on the main point. Similarly, because white exists, black can appear to the eye; there is no single truly white thing that can exist separately in the world. Having newly understood this, some people have been saying it for about fifty years. However, our Nāgārjuna and others understood precisely that in ancient times. They said also that all these external appearances do not appear outside of the projections of the mind. Whatever we see, it is seeing merely those aspects that the senses can handle, a reflection. The thing cannot be seen nakedly. Because these things were not in the least familiar to other [religions] like Christianity, scientific reasoning is considered to be something that did not exist previously. However, for us, these are familiar from long ago. Furthermore, they are amazed by the explanations in the *anuttarayoga* tantras of actually seeing the formations of the channels and drops of the body.

Yet, to be excessively proud, that is, to continually assert that even the smallest parts of all the explanations in our scriptures are unmistaken, seems beautiful only temporarily; it is a pointless stubbornness. Nothing will come from your being angry at me. If I am permitted to remain silent, I can control the peace of my own ears, but others are not benefited at all. For example, the followers of the new reasoning assert that trees are alive. Furthermore, in ancient times the Jains claimed that trees are sentient because they fold their leaves at night. [The traditional Buddhist response] was to say, "Well then, it must follow for you that pieces of leather are sentient because if they come near a fire, they shrink." It is acceptable to say, however, that there are flowers named Santeu and Venus [flytrap] that, as soon as an insect lands on them, grab it, suck its blood, hollow out the body, and discard it on the ground. Every Santeu kills more than two hundred insects every day, and the bodies just keep piling up. Similarly, in another continent, there are many trees that suck blood when they catch humans or animals. This is clear to everyone. Since these are easy to understand, I have explained them, but recently, a Bengali scholar in India invented an electronic machine that actually recognizes the presence of life. If such a flower were brought before us, would we dare contest their claim? Would we say it is the nature of the plant? Even those who assert that insects and so forth are alive must at some point show various proofs for the existence of life. Would we describe the plant as a trifling hell? However, all types of those flowers are just like that. Look at the illustration I have drawn [not extant]. The Sinhalese scientists who are Buddhists say the teacher had this in mind when he prohibited [monks from] cutting plants. But that explanation is [only] temporarily convincing.

Only fifty years ago a great debate took place between a Christian and a Buddhist in Sri Lanka. On that occasion a monk called Guṇaratna annihilated the opponents and admitted many thousands who had converted to Christianity back to Buddhism. At that time as well, none of them could refute the new reasoning. Whenever they hear talk about science, some stupid "important" people of the Tibetan race say it is the religion of Christianity. In countries that have no familiarity with Christianity, people get embarrassed and pretend [to know], when in their hearts they do not care.

I have a great desire to write a separate book on what the advantages are in considering things from the perspective of this new reasoning, but because of great difficulty and because it would become a source of disillusionment [for others], I have set the task aside.

Do not think that I am a dullard, believing immediately in whatever others say. I too am rather sharp-witted. In serving the teaching, I do not find disciples to whom I can explain the dharma. Founding a monastery requires the accumulation of many conditions. I am incapable of these great acts. [My] sympathy for the dharma is not less than yours. For that reason, do not dismiss my statements with only the wish to annihilate me. If one does not want the tree-trunk of the teachings and these roots of our Buddhist knowledge to be completely destroyed, one must be far-sighted.

Having become an open-minded person who sees the important and unimportant, you should strive to ensure the survival of the teaching [Buddhism], so that it remains together with the ways of the new reasoning. Otherwise, if, fearing complaints by others, one acts stubbornly, then one may temporarily gain great profit and many friends. As it says on the pillar at Zon de above Do-tsang [Gro tsang], "Like the light-rays of the sun and moon in the vastness of space, may the teachings of the Buddha and my reign remain equally for tens of thousands of years." Please pray that the two, this modern reasoning of science and the ancient teachings of the Buddha, may abide together for tens of thousands of years.[4]

Let us consider Gendun Chopel's brief discourse on Buddhism and Science in more detail, passage by passage.

> Now I offer a sincere discussion for those honest and far-sighted friends who are members of my religion. The system of the new reasoning "science" is spreading and increasing in all directions. In the great countries, after scattered disagreements among many people, both intelligent and stupid, saying, "It is not true," they all have become exhausted and must fall silent. In the end, even the Indian brahmans, who care more about a literal interpretation of [their] scriptures than their lives, have had to powerlessly accept it.

Gendun Chopel writes his discourse on Buddhism and Science for his fellow Buddhists back in Tibet; recognizing that it will not be immediately accepted by all who might read it, he specifies his audience as those who are "honest and far-sighted," that is, open-minded, forward-thinking, and willing to examine evidence free from prejudice. The term that he uses to translate *science* is "new reasoning" (*rigs pa gsar pa*); in the text he also transliterates the English word *science* into Tibetan as *sa yan si*, followed by the English word in capital letters and in parentheses. The term *rigs pa*, translated here as "reasoning," has a wide range of denotations in Tibet, but here seems to be used by Gendun Chopel to evoke its use in the phrase "scripture and reasoning," two traditional sources of knowledge in Buddhism, with *scripture* (*āgama* in Sanskrit, *lung* in Tibetan) referring to the discourses of the Buddha (and subsequent Indian Buddhist masters) and *reasoning* (*yukti* in Sanskrit, *rigs pa* in Tibetan) referring to the processes of logical analysis, which, in conjunction with scripture, allow one to arrive at the correct interpretation of Buddhist doctrine and, therefore, the truth. Gendun Chopel uses the latter in a broader sense here to refer to a valid way of thinking, one that is new in the sense that it does not appear in the classical presentations of reasoning, such as those of Dignāga and Dharmakīrti (but, as we shall see, compatible with them). Gendun Chopel clearly maintains the traditional sense of reasoning as a means to arrive at the truth.

As in his brief essay on the shape of the world, Gendun Chopel begins by explaining to his Tibetan audience that science is now accepted without question in "the great countries," a term of somewhat ambiguous reference, but which certainly does not include Tibet. To indicate the extent of the acceptance of science, he notes that even the brahmans in India, the epitome of hidebound conservatism and scriptural literalism, have relented.

> These assertions of the new reasoning are not established through one person arguing with another. For example, a spyglass constructed by new machines sees something thousands of miles away as if it were in the palm of one's hand, and similarly, a glass that sees what is close by makes even the smallest atoms appear the size of a mountain; one can analyze the myriad parts, actually seeing everything. Therefore, apart from closing their eyes, they [the opponents of science] had nothing else to try. At first, even those who adhered to the Christian religion in the foreign lands joined forces with the king, casting out the proponents of the new reasoning, using whatever methods to stop them, imprisoning them and burning them alive. In the end, just as one cannot hide the sunshine with one's hands, so also the parts of their religion that were unacceptable within the new system were defeated and they admitted that they were utterly false. As the glorious Dharmakīrti said [at *Pramāṇavartikka* 1.221], "Those who are mistaken about the truth cannot be changed, no matter how one tries, because their minds are prejudiced." The rejection of reason is the most despicable act.

Having described science as a new system of reasoning, Gendun Chopel wishes to make clear that it is not simply a new system of logic. Buddhist epistemology accepts two forms of valid knowledge (*pramāṇa*): direct perception (*pratyakṣa*) and inference (*anumāna*). Reasoning falls into the latter category; direct perception refers here to immediate and accurate perception through the five sense organs. Gendun Chopel wants to include direct perception in the new knowledge, especially when magnified by scientific instruments such as the telescope and the microscope. His description of a microscope as "a glass that sees what is close by [that] makes even the smallest atoms appear the size of a mountain" is presumably the first reference to this instrument to appear in Tibetan (telescopes, on the other hand, were known in Tibet); this and his description of other scientific instruments are filled with a sense of wonder, to which we will return below. What one directly perceives through these instruments is irrefutable, making it pointless to resist science.

Yet, as he explained in the essay on the shape of the world, there has been resistance in the past; early scientists were persecuted by church and state in foreign lands. His portrayal of Christianity as an opponent of science was by this time a standard trope employed by Buddhists in the Buddhism and Science discourse, and Gendun Chopel repeats it here. He goes on to explain, erroneously, that the triumph of science in the Christian nations has been so complete that those elements of Christian doctrine that are incompatible with science have all been duly abandoned. The standard corollary of this claim would be that it has not been necessary for Buddhism to renounce any of its doctrines, because Buddhism is fully compatible with Science. Buddhadasa P. Kirthisinghe wrote in 1984, "The Buddha and Buddhists welcome each scientific discovery, each new application of scientific principles, for these could never be contrary to the principles that they themselves employ."[5] Notably, Gendun Chopel does not make this claim and indeed does not hold this view, as we shall see below. He does, however, take the opportunity to cite a Buddhist text to condemn prejudice. His choice is significant.

Within the history of the Buddhism and Science discourse, it is traditional to cite passages from the Pāli canon, because it is assumed, falsely, that these are the earliest recorded statements of the Buddha and thus best represent "original Buddhism," that form of Buddhism which is both most authentic and most compatible with Science. Quotations from Mahāyāna authors are rare. But here Gendun Chopel cites the seventh-century Indian Yogācāra master Dharmakīrti. In Tibetan Buddhism, Dharmakīrti is often known simply as the "lord of reasoning" (*rigs pa'i dbang phyug*). Thus, Gendun Chopel cites the king of the old reasoning in support of the new.

> Even so, when we [Tibetans] hear the mere mention of the new system, we look wide-eyed and say, "Oh! He is a heretic!" There is the danger that when we come eventually to believe impulsively in the new reasoning, we will lose all faith in the Buddha, like some Mongolian of the Urga region [i.e., Communists], and become non-Buddhists. Therefore, whether one either stubbornly says "No!" to the new reasoning or believes in it and utterly rejects the teaching of Buddhism, both are prejudices; because it is simply recalcitrance, this will not take you far.

Unlike some other Buddhist advocates of the compatibility of Buddhism and Science, Gendun Chopel acknowledges Buddhist resistance to the scientific view, at least the resistance of Tibetan Buddhists. As is true throughout his writings, he is most willing to point out the narrow-mindedness of his fellow Tibetans, who assume that anyone who expresses the slightest interest in science must not be a Buddhist. The word translated as "heretic" is *mu stegs pa*, the Tibetan translation of *tīrthika*, the standard Buddhist term for a non-Buddhist (typically a Hindu). At the same time, he acknowledges the danger of blindly rushing to espouse science without suitable reflection and investigation; he compares such people to the inhabitants of Urga in Mongolia, once devout Buddhists, now devout Communists. His implication is that neither their Buddhism nor their Communism is authentic. He thus considers it intemperate either to reject science outright or to embrace science and reject Buddhism.

> No matter what aspect is set forth in this religion taught by our teacher [the Buddha], whether it be the nature of reality, how to progress on the path, or the good qualities of the fruition, there is absolutely no need to feel ashamed in the face of the system of science. Furthermore, any essential point [in Buddhism] can serve as a foundation for science. Among the foreigners, many scholars of science have acquired a faith in the Buddha, becoming Buddhists, and have even become monks.
>
> One person has told me: "First, I followed the system of the ancient religion of Jesus. Later, I learned science well and a new understanding was born. Then I thought that all the religions in this world are just assertions rooted in a lie, requiring that one rely only on the letter. One day, having seen a stanza of the dharma translated into a foreign language, I thought, 'Oh! The only one who follows the path of reason is the Buddha. Not only did he climb the ladder of science, but having left that [ladder] behind, he traveled even further beyond,' and conviction was born."
>
> Recently a famous renunciant named Trailokajñāna living in Sri Lanka said that in the future the religion of the Buddha will be called the religion of science, that is, the religion of reason, and other religions will be called religions of faith. Another Buddhist paṇḍita says, "Having mastered scientific reasoning, I came to believe in the Buddha. The religion of my teacher works hand in hand with scientific reasoning; when one side

tires, the other is able to leap over [to assist]. If other religions join hands with science, they collapse, either immediately or after a few steps."

Although he will qualify this statement later in the book, Gendun Chopel begins by asserting that everything set forth by the Buddha can be proudly upheld in the face of Science. In Tibetan Buddhism, the teachings of the Buddha are often divided into three categories: those on the basis (*gzhi*), those on the path (*lam*), and those on the fruition (*'bras bu*). The basis includes doctrines describing the nature of the world, on both the conventional and the ultimate levels. The path includes all those teachings of the various paths to states of liberation from suffering and rebirth. The fruition is the goal of those paths: the state of an arhat or the state of buddhahood. Everything set forth by the Buddha on these three topics is fully compatible with Science and "can serve as a foundation for science," by which he seems to mean both that it can withstand scientific investigation and that one can undertake scientific research from a Buddhist perspective.

In support of this latter point, he provides a number of testimonials, seeking to demonstrate the compatibility of Buddhism and Science by recounting the belief in this compatibility "among the foreigners." There apparently can be no greater testament than that many European scientists had become Buddhists, with some even becoming monks; people who began as Christians, then renounced all religions as superstition before eventually concluding that, from the scientific perspective, Buddhism is better than atheism. Just as the Buddha had used the raft of the dharma to reach the other shore, then left the raft behind, this unidentified Buddhist scholar had used the ladder of science to transcend science.

It would have been interesting if Gendun Chopel had provided the names of these European scientists who converted to Buddhism. He does mention one name, however. He refers to the "famous renunciant" Trailokajñāna, his Sanskrit approximation of Ñāṇatiloka (also spelled Nyanatiloka). This was the Buddhist name of Anton Walter Florus Gueth (1878–1957), a violinist who was the first German to be ordained as a Buddhist monk and who went on to become a revered and eminent scholar of Theravāda Buddhism, living in an island hermitage in Sri Lanka. It is unclear whether Gendun Chopel met Ñāṇatiloka during his

1940–41 visit to Sri Lanka; as a German national in the British colony of Ceylon, Ñāṇatiloka was interned in India by the British during the Second World War. But Gendun Chopel certainly knew about him, and quotes his prediction that Buddhism will become the religion of Science. He quotes someone else to the effect that among all the religions, only Buddhism is able to work hand in hand with Science. Gendun Chopel next turns to some examples of the compatibility of Buddhism and Science.

> For example, the followers of the new reasoning say, "In the second moment immediately after any object comes into existence, it ceases or dissolves. These collections of disparate things disperse like lightning." Consequently, the first moment of a pot does not persist to the second moment, and even the perception of a shape does not exist objectively apart from the power of mind or of human language. Moreover, when examined as above, even colors are merely the ways a wave of the most subtle particles moves. For example, regarding waves of light, there is no difference of color whatsoever to be seen in the particles that are the basis for that color; it is simply that eight hundred wavelengths in the blink of an eye appear as red and four hundred appear as yellow. Furthermore, they have invented another apparatus for seeing things that move too quickly to be seen, like drops of falling water. Something that lasts for one blink of an eye can be easily viewed over the duration of six blinks of an eye. More than ten years have passed since they made a viewing apparatus that is not obstructed [in seeing] things behind a wall or inside of a body. All of this is certain. They have also made a machine by which what is said in India can be heard in China in the following moment. Because they are able to show in China a film of something that exists in India, all people can be convinced. The final proof that all things run on waves of electricity is seeing it with one's own eyes.

Having explained that European scientists believe that the teachings of the Buddha are fully compatible with Science, here Gendun Chopel turns to the description of some of their discoveries. The first examples are presented in the language of Buddhist philosophy, immediately comprehensible to any Tibetan scholar who would read his words. The European scientists assert that an object ceases to exist in the second moment after it comes into existence. This seems identical to the Buddhist doctrine of subtle impermanence, according to which all conditioned things go through a process of production, abiding, aging, and disintegration in each moment. Gendun Chopel was aware of the disagreements between the Sarvāstivāda and Sautrāntika schools of Indian Buddhism on this point, as to whether production, abiding, aging, and disintegration were qualities that acted upon an object or whether they were instead descriptions of processes inherent to the object itself. His statement that "even the perception of a shape does not exist objectively apart from the power of mind or of human language" evokes the Yogācāra doctrine (found elsewhere, although less strongly, in other schools of Buddhist philosophy) that an external object does not exist separately from the consciousness that perceives it, and that objects exist as designations by human thought and language; objects are not naturally the bases of their names. European scientists had apparently come to these conclusions through observation, enhanced by instruments unknown in Tibet. Next, describing what appears to be the use of a spectrophotometer, Gendun Chopel explains that colors are in fact wavelengths of light moving at different speeds. Here he shifts abruptly into a vocabulary that would be quite foreign to a Tibetan reader, even one with a knowledge of Buddhist epistemology, where this kind of analysis of color is not to be found. Yet the discovery that light consists of rapidly moving waves would be regarded as generally consistent with the doctrine of impermanence. He next seems to describe slow motion photography that allows the perception of things that move too quickly for the naked eye. His Tibetan readers would be reminded of the Buddhist claim that subtle impermanence can only be inferred by ordinary beings but can be directly perceived by an enlightened being.

But Gendun Chopel's reference to spectrometry and slow-motion photography seems to remind him of even more wondrous things that he has heard about, things that do not fit so easily into the categories of Buddhist epistemology. The viewing apparatus that can see through walls is apparently an X-ray machine; the machine that causes something said in India to be heard in China in the next moment is presumably a radio. The ability to see things obstructed by other objects and to hear at great distances are standard supernormal powers (*abhijñā*) in Buddhism, the by-products of the achievement of deep states of concentration (and thus available to Buddhist and non-Buddhist yogins alike). But these do not seem to be his referents here. Instead, he speaks of the power of electricity to expand the capacity of direct perception, to hear in China what was said in India by simply listening to a radio; to see in China what has happened in India by watching a film. Radio and motion pictures for Gendun Chopel are not media but extensions of sense perception. Electricity (*glog*, literally "lightning" in Tibetan), which had been introduced in Lhasa on a limited basis in the 1930s, was not a modern wonder but rather the fundamental energy of the universe. Quickly returning to his topic, Gendun Chopel identifies the Buddhist analogs to the scientific discoveries, analogs to which he had only alluded above.

> Many great scholars of science made limitless praises of the Buddha, saying that two thousand years ago, when there were no such machines, the Buddha explained that all compounded things are destroyed in each moment and he taught that things do not remain even for a brief instant, and subsequently we have actually seen this using machines. The statement by Dharmakīrti that "continuity and collection do not exist ultimately" can be understood in various ways, but in the end one can put one's finger on the main point. Similarly, because white exists, black can appear to the eye; there is no single truly white thing that can exist separately in the world. Having newly understood this, some people have been saying it for about fifty years. However, our Nāgārjuna and others understood precisely that in ancient times. They said also that all these external appearances do not appear outside of the projections of the mind. Whatever we see, it is seeing merely those aspects that the senses can handle, a reflection. The thing cannot be seen nakedly. Because these things were not in the least familiar to other [religions] like Christianity, scientific reasoning is considered to be something that did not exist previously. However, for us, these are familiar from long ago. Furthermore, they are amazed by the explanations in the *anuttarayoga* tantras of actually seeing the formations of the channels and drops of the body.

That science eventually reveals what the Buddha naturally knew is a standard element of the Buddhism and Science discourse. Gendun Chopel provides his list of examples here, ones that draw upon the scholastic training of a Tibetan monk. Modern machines have confirmed the Buddha's ancient description of subtle impermanence. Although the statement of the great Buddhist logician Dharmakīrti that "continuity and collection do not exist ultimately" is open to a wide range of interpretation, no one would dispute that it also means that the autonomy of objects in time and space is only apparent; they are in fact impermanent processes coming into and going out of existence in each moment. The precise differentiation of the number of wavelengths that compose a given color was unknown to the Buddhists. In a somewhat forced analogy, Gendun Chopel notes that the great second-century Indian Madhyamaka master Nāgārjuna explained that nothing exists in and of itself, everything exists in dependence on something else, such that there is black only because there is white. All the major schools of Indian Buddhist philosophy (except for one, the Vaibhāṣika) argued that the senses did not perceive their objects directly, but rather perceived an "aspect" (*ākāra*) of the object, like a reflection cast by the object. All of these things have been long familiar to Buddhists, although scientists regard them as discoveries.

As noted above, although Gendun Chopel's general point is a familiar one in the Buddhism and Science discourse, the specific examples he provides are unusual, especially for his time, drawing as he does largely on Mahāyāna figures like Nāgārjuna and Dharmakīrti. Most unusual, however, is his brief allusion here to tantra: "Furthermore, they are amazed by the explanations in the *anuttarayoga* tantras of actually seeing the formations of the channels and drops of the body." At the time that Gendun Chopel wrote these words, Buddhist tantra was widely regarded by European scholars as a late accretion of magic and superstition into the Buddhist tradition, utterly unconnected to the teachings of the Buddha himself. Gendun Chopel held a very different view, seeing the tantras, and especially the *anuttarayoga* tantras, the "highest yoga tantras," as the supreme and secret teachings of the Buddha, indispensable to the achievement of buddhahood. According to the *anuttarayoga* tantra systems, buddhahood is achieved through the practice of the "stage of generation" and the "stage of completion." In the latter, "drops," concentrations of energy, are caused to move through a system of seventy-two thousand subtle channels that radiate through the body, bringing about deep states of bliss and insight. This is what Gendun Chopel is alluding to; he seems to imply that the tantras were describing the central nervous system centuries ago. However, when he wrote these words (presumably in 1940 or 1941), this tantric physiology was not widely known in Europe and America, making it unclear which explanations of it scientists might have heard. Given the paucity of scholarship on Buddhist tantra at that time, and the low regard in which it was held, it is possible that here Gendun Chopel is reporting his own conversations on the topic. Regardless, this sentence is perhaps the first reference to Buddhist tantra in the Buddhism and Science discourse. It would not be the last.

This reference to tantra is not Gendun Chopel's only deviation from the genre. Buddhist apologists have traditionally been loath to concede any point on which Buddhist doctrine may be found in error in light of scientific knowledge; Gendun Chopel himself sought to maintain that the Buddha knew that the world is round, although he said it is flat. But in the next passage we find a concession, and a caution.

> Yet, to be excessively proud, that is, to continually assert that even the smallest parts of all the explanations in our scriptures are unmistaken, seems beautiful only temporarily; it is a pointless stubbornness. Nothing will come from your being angry at me. If I am permitted to remain silent, I can control the peace of my own ears, but others are not benefited at all. For example, the followers of the new reasoning assert that trees are alive. Furthermore, in ancient times the Jains claimed that trees are sentient because they fold their leaves at night. [The traditional Buddhist response] was to say, "Well then, it must follow for you that pieces of leather are sentient because if they come near a fire, they shrink." It is acceptable to say, however, that there are flowers named Santeu and Venus [flytrap] that, as soon as an insect lands on them, grab it, suck its blood, hollow out the body, and discard it on the ground. Every Santeu kills more than two hundred insects every day, and the bodies just keep piling up. Similarly, in another continent, there are many trees that suck blood when they catch humans or animals. This is clear to everyone. Since these are easy to understand, I have explained them, but recently, a Bengali scholar in India invented an electronic machine that actually recognizes the presence of life. If such a flower were brought before us, would we dare contest their claim? Would we say it is the nature of the plant? Even those who assert that insects and so forth are alive must at some point show various proofs for the existence of life. Would we describe the plant as a trifling hell? However, all types of those flowers are just like that. Look at the illustration I have drawn [not extant]. The Sinhalese scientists who are Buddhists say the teacher had this in mind when he prohibited [monks from] cutting plants. But that explanation is [only] temporarily convincing.

He begins with the concession, relatively rare in the genre, that it is short-sighted and stubborn to hold that every point of Buddhist doctrine is unmistaken and confirmed by modern science. Recognizing that such a statement might seem heretical to his readers, he asks for their indulgence. He obviously feels that what he has to say is important for Tibetans to hear. If he remains silent, he can preserve the peace of his ears, because he will not have to listen to the condemnations that his words will provoke. But if he remains silent, he will not be heard, and it is his conviction that his words are important. He then begins a discussion of an important point on which Buddhism and Science disagree: the sentience of plants.

This question also was a point of disagreement between the Buddhists and the Jains in ancient India. The Buddhists argued that there were six, and only six, types of sentient beings, and hence six possibilities for rebirth: gods, demigods, humans, animals, ghosts, and hell beings. Plants and inanimate objects did not have consciousness, and thus one could not be reborn as a plant or inanimate object. The Jains posited the existence of beings that had one sense organ (*ekendriya*), the sense of touch, which included plants, stones, and water. Among the Jain evidence for the sentience of plants was that some plants fold their leaves at night. The Buddhists would counter with the absurd consequence that a piece of leather must be sentient because, when it is placed near a fire, it draws back.

Gendun Chopel was known as a skilled debater during his days in the monastery. One of the marks by which this proficiency was measured in Tibet was the ability to successfully defend a non-Buddhist (and hence wrong) position against the Buddhist (and hence correct) position. Although he does not mention it here, Gendun Chopel was famous in his youth for having successfully upheld the Jain view that plants have consciousness. Arriving in India many years later, he seems to have found empirical proof. He consequently describes the Venus flytrap and another carnivorous plant called Santeu (perhaps the Nepenthes or "pitcher plant" found in Sri Lanka) and also provided a drawing, which apparently has been lost. His claim that, "in another continent, there are many trees that suck blood when they catch humans or animals" seems to have passed over into the realm of the fantastic. What he refers to when he describes "a machine that actually recognizes the presence of life" is unclear.

Gendun Chopel's understanding of botany was such that he concluded that a Venus flytrap has consciousness because it consumes insects. For him, this is clear empirical evidence that the Buddhist view regarding the consciousness of plants is mistaken. His question, then, is how Buddhists should respond. One might concede that it is indeed the nature of the plant, and that the plant is sentient. There is also a doctrinal loophole available to the Buddhist. Among the various forms of hell described in the Buddhist cosmographies are the trifling (*prādeśika*) hells. Within these categories are beings temporarily born inside inanimate objects (rocks and brooms are sometimes mentioned) and who mistakenly identify such objects as their bodies. Thus, Gendun Chopel asks somewhat sarcastically whether the Buddhist might wish to uphold the position that plants lack consciousness by saying that a Venus flytrap is a case of a trifling hell. But this would not be an adequate response, because in his view all plants have consciousness.

Gendun Chopel mentions that Sinhalese scientists (who are also Buddhists) acknowledge that plants have consciousness and argue that the Buddha knew this because he prohibited monks from damaging plants. It is the case that damaging a living plant is one of the minor violations (to be expiated through confession) of the monastic code. The Buddha is said to have made this rule, however, because plants (especially trees) are sometimes the abodes of local spirits; the term translated as "plant" is *bhūtagāma*, "abode of a being." Gendun Chopel seems to regard such an argument as special pleading, calling it "temporarily convincing."

> Only fifty years ago a great debate took place between a Christian and a Buddhist in Sri Lanka. On that occasion a monk called Guṇaratna annihilated the opponents and admitted many thousands who had converted to Christianity back to Buddhism. At that time as well, none of them could refute the new knowledge. Whenever they hear talk about science, some stupid "important" people of the Tibetan race say it is the religion of Christianity. In countries that have no familiarity with Christianity, people get embarrassed and pretend [to know], when in their hearts they do not care.

Here Gendun Chopel alludes to the famous debate between a Sri Lankan monk and a Sri Lankan Christian clergyman. However, he is mistaken concerning the particulars. The debate took place at Pānadurē in 1873, twenty years earlier than Gendun said. The name of the Buddhist speaker was Guṇānanda (Migettuwatte Guṇānanda Thera, 1823–1890), not Guṇaratna. The debate is discussed in some detail in chapter 1, and there is no need to repeat that discussion here. Yet it is noteworthy that Gendun Chopel takes great interest in the Pānadurē debate, mentioning it at several points in his writings. In his description of it elsewhere in *The Golden Chronicle*, he writes, "They did not debate about the essential points of the [Buddhist] view [that there is no self]. Rather, the debates by the opponents were based on a knowledge of modern science. Thus, I think it would have been extremely difficult for people like us [Tibetans] to respond."[6]

There seem to be a number of reasons for his interest in the event. First, he takes a certain pride in the defeat of the Christians by the Buddhists, especially in the forum of a debate, for which there is a long and storied tradition in Tibet. Second, as the passage just quoted indicates, the Buddhists did not defeat the Christians in a debate on the validity of points of Buddhist doctrine, although there are numerous instances of such victories over Hindu opponents in the history of Indian Buddhism. Instead, at least as Gendun Chopel reports it, the debate centered on Buddhist and Christian understandings of science. That a Buddhist was able to defeat a Christian in this arena, despite the association of science with Christian Europe, is yet further proof that Buddhism surpasses all religions in its compatibility with Science. Finally, the debate provides him with yet another opportunity to tweak his Tibetan readers. Some ignorant Tibetan aristocrats and officials identify science with Christianity, but the Pānadurē debate, in which a Buddhist monk defeated a Christian minister in battle over knowledge of science, proves that this is not the case. In fact, Tibetans have little familiarity with Christianity and only pretend to understand it. The debate also proves how advanced the Buddhists of Sri Lanka are in comparison with those of Tibet. If a Tibetan monk had to engage in a debate about science, Gendun Chopel thinks "it would have been extremely difficult for people like us to respond."

> I have a great desire to write a separate book on what the advantages are in considering things from the perspective of this new reasoning, but because of great difficulty and because it would become a source of disillusionment [for others], I have set the task aside.

> Do not think that I am a dullard, believing immediately in whatever others say. I too am rather sharp-witted. In serving the teaching, I do not find disciples to whom I can explain the dharma. Founding a monastery requires the accumulation of many conditions. I am incapable of these great acts. [My] sympathy for the dharma is not less than yours. For that reason, do not dismiss my statements with only the wish to annihilate me. If one does not want the tree-trunk of the teachings and these roots of our Buddhist knowledge to be completely destroyed, one must be far-sighted.

During his time in South Asia, Gendun Chopel was obsessed with his legacy, fearing that his words would not be heard by his countrymen—or, if they were heard, that they would be rejected. He was learning many things during his years abroad, things that no other Tibetan had learned. Not yet forty years old, he perhaps had a somewhat inflated view of his own importance, and imagined that Tibetans would receive his work more seriously (regardless of the response) than they did. The sad fate that he met upon his return to Lhasa (the precise reasons for which remain unclear), however, also suggests that his iconoclasm came at a price.

In the passage above, he renounces his wish to write an entire book devoted to science, for fear that it would become a source of disillusionment to others, by which he would seem to mean that they would lose their faith in Buddhism. This is a somewhat surprising sentiment, given his conviction of the compatibility of Buddhism and Science. Indeed, he goes on to say that he is prevented by circumstance from performing traditional deeds in service of the Buddha's teaching, such as teaching the dharma to a large number of disciples or founding a great monastery. Yet his devotion to the dharma is no less than that of others. He thus asks that his discourse on Buddhism and Science not be rejected outright, but that it be seen instead as his contribution to the survival of Buddhism into the modern age.

> Having become an open-minded person who sees the important and unimportant, you should strive to ensure the survival of the teaching [Buddhism], so that it remains together with the ways of the new reasoning. Otherwise, if, fearing complaints by others, one acts stubbornly, then one may temporarily gain great profit and many friends. As it says on the pillar at Zon de above Do-tsang [Gro tsang], "Like the light-rays of the sun

> and moon in the vastness of space, may the teachings of the Buddha and my reign remain equally for tens of thousands of years." Please pray that the two, this modern reasoning of science and the ancient teachings of the Buddha, may abide together for tens of thousands of years.

Gendun Chopel concludes his discourse on science with an exhortation. He is convinced that Buddhism can survive only as the ally of science, not as its opponent. Condemning science as inimical to Buddhism may garner temporary fame, but it is ultimately shortsighted and narrow-minded. He asks his fellow Tibetans to pray that Buddhism and Science may flourish together for many millennia. That is, he offers a Buddhist prayer for science, adapting the words carved on a stone pillar in Tibet in the distant past to pray for the compatibility of Buddhism and Science far into the distant future.

We see in Gendun Chopel's writings on science a clear, and rather unrestrained, sense of wonder at the marvels of modern technology. But that wonder is then tempered by claims of the Buddha's anticipation of the recent discoveries of the Europeans. Lest this sense of wonder at the apparently miraculous be mistaken for the simple open-mouthed awe of the naïve native, Gendun Chopel conjoins it with a sense of recognition of the familiar. Upon reflection, the Buddhist, or at least the Buddha, has seen this all before.

Yet in another work, a poem sent back to his boyhood monastery in northeastern Tibet, Gendun Chopel succumbs to the sensational. In addition to its unabashed incredulity, the poem offers an unexpected insight into the history of the discourse of Buddhism and Science.

> A small octagonal box with a small wheel
> When spun by just one man
> Can tremble the earth twenty *yojana*s away.
> That seems like a lie. Isn't it amazing?
> Oh, that's not so amazing.
>
> By removing a small scar
> An old man of over eighty years
> Can be made twenty years younger.
> That seems like a lie. Isn't it amazing?
> Oh, that's not so amazing.

Through the emptiness of outer space
A lightning letter with great inner speed
Can travel ten thousand *yojana*s.
That seems like a lie. Isn't it amazing?
Oh, that's not so amazing.

A beam of white light narrow as a single hair of a horse's tail
When sent from a far away place
Can control a machine of great power.
That seems like a lie. Isn't it amazing?
Oh, that's not so amazing.[7]

The poem is undated; it was certainly written during his sojourn in South Asia, therefore between 1934 and the end of 1945. The fact that he does not refer to any weapons of war suggests that it might have been written before the outbreak of the Second World War. The first stanza seems to depict some kind of explosion triggered by remote control. A *yojana* is the standard measure of distance in classical Indian literature; unfortunately, its length is not standard, and is typically estimated as being between 5 and 10 miles. Gendun Chopel seems to be describing a device that sends a spark to ignite dynamite, but such a device would require a wire connecting the box to the explosive, and that wire would need to be between 50 and 100 miles long. The second stanza describes a surgical procedure that can make an old man younger, or perhaps only appear younger. Cosmetic surgery was not practiced at this time, at least in India. The term translated as "scar" (*sha rmen*) may refer to a gland, but even then the reference is not obvious. The third stanza refers to a "lightning letter" (*glog gi 'phrin*), the Tibetan term for a telegram. Gendun Chopel's sense of amazement causes him to exaggerate distances, claiming that the telegram can travel 10,000 *yojana*s, a distance somewhere between 50,000 and 100,000 miles; the circumference of the earth at the equator is 24,901.55 miles. The final stanza seems to repeat the earlier reference to some version of remote control.

Gendun Chopel thus presents four examples of the wonders of science, in the form of a Tibetan version of "Believe It or Not." What is most striking, however, is that just sixty or seventy years after he composed this poem, the technology it describes is so arcane as to be unrecognizable. The sole exception is the telegram, an invention that has since been consigned to the science museum. Gendun Chopel obviously had something in mind, something that he had seen or, more likely, heard or read about in each case. But his descriptions are sufficiently vague as to obscure his references; his understanding of the technology was limited at best, and likely wrong. This raises the question of what *Science* means to the representative of *Buddhism*. It also raises this question: if Buddhism, however it is understood, was somehow compatible with the science of the first half of the twentieth century, science that has been superseded to the point that its description by a Buddhist of the day is unrecognizable, how can Buddhism also be compatible with the science of the second half of the twentieth century—or the science of the first half of the twenty-first?

And as science changes, does Buddhism, renowned for its doctrine of impermanence, remain ever the same? Historians have duly documented the various transformations of Buddhist thought that have occurred across Asia over many centuries. But from the Buddhist perspective, the Buddhist truth is timeless; the Buddha understood the nature of reality fully at the moment of his enlightenment, and nothing beyond that reality has been discovered since, because nothing beyond that reality exists. Thus, in order to maintain the claim that Buddhism is compatible with Science, as Science continues to change, it has been necessary for Buddhist thinkers to change the unchanging truth said to be at the heart of Buddhism. From one perspective, this is exactly how Buddhist thought has developed, with monks across Asia refining and redefining the tradition, but with the constant claim that they are engaged always in recovery, never in innovation. All of the innovations in the history of Buddhist thought have been presented as reclamations of the content of the Buddha's enlightenment.

Awe can become awful; fascination can turn to fear. The scientific wonders that Gendun Chopel and his Buddhist brethren saw, heard, and read about, despite their hypothetical anticipation by the Buddha of Asia, were in fact all products of foreigners. Fear of the foreign is perhaps endemic, as the encounter with the unfamiliar breeds misapprehension and suspicion. Gendun Chopel relates elsewhere that when the Portuguese arrived in Sri Lanka, the Sinhalese reported that the foreigners ate rocks and drank blood; he speculates that they saw them eating bread and drinking tea. And the foreigners who brought science to South Asia were not benign. They were colonizers, and Gendun Chopel speaks scathingly of the British elsewhere in his writings. Describing the first wave of European colonialism, he writes, "Sponsored by kings and ministers who disregarded others' welfare, they trod upon the happiness of others like a turnip on the ground, sending out a great army of bandits, calling them 'traders.'"[8] Describing the British India he encountered in the 1930s, he writes:

> They introduce the new ways of modernity, such as railroads, schools, and factories. Their law is only good for the educated and the wealthy families. If one has money and education, anything is permitted. As for the lowly, their small livelihoods that provide the necessities for life are sucked like blood from all their orifices. Such a wondrous land as India appears to be filled with poor people who are like hungry ghosts.[9]

His sense of wonder and his suspicion come together in this short poem about a motion picture.

> On a curtain of stainless white silk
> Filled with magical electric light
> Is the Queen of Illusion, laughing and crying,
> Making a show for visitors of the three realms.
>
> Lacking the oil of compassion that benefits others,
> Those skilled in the arts, with the sorcery of electricity,
> Show the crooked path to honest humans.
> Beware the race of yellow-haired monkeys.[10]

White silk connotes purity and honesty in Tibetan culture, with the best silk said to come from Banaras. This Asian silk is then pervaded by light from the foreigner's movie projector, to reveal the Queen of Illusion (*sgyu ma'i rgyal mo*). The word *illusion* here has strong connotations of deception, magic, and trickery, of an appearance at odds with reality. Her display of emotions captivates the audience of "visitors of the three realms," that is, the beings who wander in through the realms of rebirth. Thus, into the world of saṃsāra, the magic of electricity projects another layer of illusion.

The second stanza describes the projectionist. Calling Europeans "the race of yellow-haired monkeys" (a highly pejorative phrase in Tibetan),

Gendun Chopel finds them to be merely clever, rather than wise, possessed of apparently miraculous power, but lacking compassion; they are skilled in the art of deception. They mislead those who are by nature honest; the third line reads more literally, "show the crooked path to straight humans."

Gendun Chopel's poem is yet another reminder that the discourse of Buddhism and Science is the product of the colonial encounter. When the Pānadurē debate took place in Sri Lanka in 1873, Sri Lanka and India were British colonies. When Gendun Chopel wrote his poem, circa 1940, Sri Lanka and India were British colonies. The proponents of Buddhism's compatibility with Science, Guṇānanda and later Dharmapāla in Sri Lanka, Shaku Sōen (among many others) in Japan, and Gendun Chopel in India, were Buddhist apologists—not in the colloquial sense of having something to apologize for, but in the classical sense of *apologia*, a Greek legal term referring to the formal defense presented in reply to the charges of the prosecution. The prosecutors in the case of Buddhism were Christian missionaries (from Europe and America) and modern secularists (especially in Japan and China) who saw Buddhism as an outmoded superstition. It was the task of the Buddhist apologists to counter this claim by demonstrating that in fact Buddhism is rational, scientific, and modern, indeed more rational, scientific, and modern than Christianity (and all other religions) and thus best suited for the modern world. But the age of colonialism seems long past. Why continue to make these claims long after the battle has been won?

• • •

The Fourteenth Dalai Lama was born in 1935, the year after Gendun Chopel's departure for India. Although Gendun Chopel spent the last five years of his life in Lhasa, he would never meet the Fourteenth Dalai Lama. Three of those last five years were spent in prison; according to some reports, Gendun Chopel was released as part of a general amnesty to celebrate the Dalai Lama's birthday. Thus, Gendun Chopel never saw the Dalai Lama, but the Dalai Lama may have seen him. The prison where Gendun Chopel was incarcerated was located at the very foot of the Potala, the Dalai Lama's winter palace, and the young Dalai Lama may have observed him from afar, with the telescope through which he looked down upon his subjects; the Dalai Lama is regarded as the human incarnation of the bodhisattva of compassion, Avalokiteśvara, whose name means "the Lord Who Looks Down" in Sanskrit.

Yet the two had some things in common. They were both from Amdo, the far northeastern region of the Tibetan cultural domain. They were both Buddhist monks of the same sect, trained at the highest level of the monastic academy. Most important for our purposes, they were the two most prominent Tibetan Buddhist thinkers of the twentieth century to take an active interest in science. Although their writings on the subject are separated by more than half a century, their views are remarkably similar.

The Dalai Lama is the author of dozens of books. Only a few, such as his first autobiography, *My Land and My People* (1962); his first book on Buddhism, *Opening the Eye of New Awareness* (published in Tibetan in 1963); and essays such as *The Buddhism of Tibet and the Key to the Middle Way* (1975), were written originally in Tibetan. The great majority of the publications are translated and edited transcriptions of the many series of lectures on Buddhist topics that the Dalai Lama has delivered in Tibetan; some are edited transcripts of conversations and interviews with the Dalai Lama, conducted in Tibetan and English. A few books offer a sustained presentation of the Dalai Lama's views on a specific topic (in contrast to a commentary on a Buddhist text or responses prompted by an interviewer's questions), such as *Ethics for a New Millennium* (1999) and *The Universe in a Single Atom* (2005). The Dalai Lama has discussed science in many of his interviews and lectures, especially over the past two decades. However, this last book, subtitled *The Convergence of Science and Spirituality*, offers both his most recent and his most sustained discussion of Buddhism and Science, and thus shall be our focus here.[11]

The title of the book is drawn not from William Blake but from a famous passage in the *Garland Sūtra* (*Avataṃsaka* or *Buddhāvataṃsaka*), an important Mahāyāna sūtra, portions of which date from the first century CE. The passage reads, "In each atom of the realms of the universe, there exist vast oceans of world systems." This statement is especially famous in Chinese Buddhism, particularly in the Huayan school, although the same point is made repeatedly in Mahāyāna sūtras.

In the prologue, the Dalai Lama states:

> Many years after I went into exile in India, I came across an open letter from the 1940s addressed to the Buddhist thinkers of Tibet. It was written

> by Gendün Chöphel, a Tibetan scholar who not only had mastered Sanskrit but also, uniquely among Tibetan thinkers of his time, had a good command of English. He traveled extensively in British India, Afghanistan, Nepal, and Sri Lanka in the 1930s. This letter, composed toward the end of his twelve-year trip, was amazing to me. It articulates many of the areas in which there could be fruitful dialogue between Buddhism and modern science. I discovered that Gendün Chöphel's observations often coincide remarkably with my own. It is a pity that this letter did not attract the attention it deserved, partly because it was never properly published in Tibet before I came into exile in 1959. But I find it heartwarming that my journey into the scientific world has a precedent within my own Tibetan tradition. All the more so since Gendün Chöphel came from my native province of Amdo. Encountering this letter so many years after it was written was an impressive moment.[12]

The "open letter" to which the Dalai Lama refers is undoubtedly the passage from the final chapter of *The Golden Chronicle* that we have just considered. Probably composed in 1940 or 1941, it would not be published until 1990.

In *The Universe in a Single Atom*, when the Dalai Lama speaks of Buddhism, he is most often referring to Mahayana Buddhist philosophy as it is understood in Tibet. Thus, the Madhyamaka master Nāgārjuna and the Yogācāra master Dharmakīrti, the key figures of what might be called Buddhist ontology and logic, respectively, are mentioned most often, more than the Buddha himself. Such important Mahāyāna thinkers as Asaṅga, Vasubandhu, and Bhāvaviveka also appear. The tenets of the Hīnayāna Abhidharma are also discussed, but often as examples of the Buddhist views that must be dismissed in light of the discoveries of science. In some ways, this is most traditional. In the Dalai Lama's Geluk sect, it is asserted that the "higher" Mahāyāna schools of Yogācāra and Madhyamaka are able to refute the positions of the "lower" Hīnayāna schools. And one of the hallmarks of the thought of Tsong kha pa (1357–1419), the founder of the Geluk, was that the highest reality, the emptiness of intrinsic existence described by Nāgārjuna, must first be approached through the logical methods set forth by Dharmakīrti. It is noteworthy that Gendun Chopel, writing a half century before, evokes the same figures. Indeed, in the discourse of Buddhism and Science, "Buddhism" is typically represented as the philosophical position of the school of Buddhism of the author, whether that be the Abhidharma in the case of Theravāda or Madhyamaka in the case of Mahāyāna. What Tibetan thinkers like Gendun Chopel and the Dalai Lama add that has traditionally been absent from the discourse is tantra; Gendun Chopel mentions the *anuttarayoga* tantras, and the Dalai Lama alludes to the *Kālacakra*.

The Universe in a Single Atom recounts the Dalai Lama's first encounters with science as a young boy in Tibet, where the German mountaineer Heinrich Harrer taught him about geography. He describes his tour of Mao's China in 1954, where he saw hydroelectric dams, and his visit to India in 1956, where he had his "first encounters with spiritual teachers who were seeking the integration of science and spirituality, such as the members of the Theosophical Society in Madras."[13] He discusses his conversations with a number of prominent scientists and thinkers over the past thirty years, describing, in a self-effacing way, what he learned from them. He intersperses these accounts with stories from his youth in Tibet and explanations of various salient doctrines in the Buddhist tradition. The Dalai Lama discusses (in order) physics, cosmology, evolution, consciousness, and genetics, concluding with a chapter entitled "Science, Spirituality, and Humanity." His general thesis is stated clearly in the prologue.

> The great benefit of science is that it can contribute tremendously to the alleviation of suffering at the physical level, but it is only through the cultivation of the qualities of the human heart and the transformation of our attitudes that we can begin to address and overcome mental suffering. In other words, the enhancement of fundamental human values is indispensable to our basic quest for happiness. Therefore, from the perspective of human well-being, science and spirituality are not unrelated. We need both, since the alleviation of suffering must take place at both the physical and psychological levels.[14]

This is an important statement, clearly predicated on the Buddha's four noble truths, said to have been set forth in his first sermon after his enlightenment: that life is qualified by suffering and that the most pressing task is to alleviate that suffering. Suffering can take two forms, physical and mental; and science, in the form of modern medicine, has made great strides in reducing physical suffering. As other Buddhist thinkers have argued in the past, the Dalai Lama suggests that science only proves effective against physical suffering. Mental or "psychological" suffering must be countered with spirituality. (The Dalai Lama consistently uses *spirituality* rather than *religion* throughout the book. Although the distinction is not explained, he presumably is using the former to refer to any number of contemplative traditions, both new and old, including, of course, Buddhism. The term *spirituality* does not have an obvious correlate in Tibetan.) That is, Science understands the outer world, Buddhism understands the inner world. This is a persistent trope in the discourse of East and West, Buddhism and Science.

Buddhism has also explained the outer world over the course of its long history, and in the view of the Dalai Lama, several of these teachings are simply wrong. As noted in the discussion of the Mount Meru cosmology in chapter 1, he is willing to dismiss a number of Buddhist doctrines about the physical universe. There I cited his statement in *The Way to Freedom*, where he wrote, "The purpose of the Buddha coming to this world was not to measure the circumference of the world and the distance between the earth and the moon, but rather to teach the Dharma, to liberate sentient beings, to relieve sentient beings of their sufferings."[15] In *The Universe in a Single Atom*, he states that "if scientific analysis were conclusively to demonstrate certain claims in Buddhism to be false, then we must accept the findings of science and abandon those claims."[16] The important phrase here is "conclusively demonstrate to be false." Thus, photographs of the earth taken from space conclusively demonstrate that the world is not flat, as the sūtras describe it. However, the Dalai Lama makes a distinction, drawn from Madhyamaka philosophy, between not finding something (*ma snyed pa*) and determining that it does not exist (*med pa nges pa*). For him, the flat earth has been determined not to exist. But other Buddhist doctrines, most important, rebirth, have simply not been found. Indeed, a central concern of *The Universe in a Single Atom* is to defend one Buddhist doctrine (rebirth) and one Buddhist value (compassion) against possible scientific refutation.

These issues are not raised until the end of the book. In the earlier chapters, the Dalai Lama takes up other topics that have become standard elements of the Buddhism and Science discourse. As noted earlier, Gendun Chopel had extolled Nāgārjuna's prescience in declaring the relative nature of all phenomena many centuries ago. The Dalai Lama writes, "To a Mahayana Buddhist exposed to Nagarjuna's thought, there is an unmistakable resonance between the notion of emptiness and the new physics. If on the quantum level, matter is revealed as less solid and definable than it appears, then it seems to me that science is coming closer to the Buddhist contemplative insights of emptiness and interdependence."[17] Here he notes only the resonance of emptiness and quantum physics, but implies that there is a single truth (in this case, emptiness) discovered by Buddhists millennia ago, which modern science is only coming to see. He also describes the relation of emptiness and interdependence (*pratītyasamputpāda*, often translated as "dependent origination" or "dependent arising") as a contemplative insight.

For the Dalai Lama, these resonances are possible, in part, because Buddhism and Science are both true, when "Buddhism" is freed from cultural accretions, such as the flat-earth theory. But they are also possible because of a basic similarity between Buddhism and Science, one which has been proclaimed by Buddhists since the days when Science described a mechanistic universe. "Buddhism and science share a fundamental reluctance to postulate a transcendent being as the origin of all things. This is hardly surprising given that both these investigative traditions are essentially nontheistic in their philosophical orientations."[18]

In keeping with the view of Buddhism as an "investigative tradition," and in harmony with those who, over the past century, have claimed that Buddhism and Science share a similar method, the Dalai Lama writes, "So one fundamental attitude shared by Buddhism and science is the commitment to keep searching for reality by empirical means and to be willing to discard accepted or long-held positions if our search finds that the truth is different."[19] Setting aside the question of "empirical means" for the moment, one might first consider the issue of the search for reality in Science and in Buddhism. In the scientific method, at least in its idealized form, reality or truth (whatever those terms might mean in a given case) has not yet been discovered; hypothesis and experiment are employed to arrive at a truth that is at the time unknown, or at least unverified. In the various Buddhist traditions, there is the shared belief that the nature of reality was discovered long ago by the Buddha, and before him by the buddhas of the distant past. Reality is represented not as something that the Buddha was the first to discover but as that which he revealed, the ancient city at the end of the ancient path through the great forest. In a famous passage in the *Saṃyutta Nikāya*, the Buddha says:

> Suppose, monks, a man wandering through a forest would see an ancient path, an ancient road traveled upon by people in the past. He would follow it and would see an ancient city, an ancient capital that had been inhabited by people in the past, with parks, groves, ponds, and ramparts, a delightful place. Then the man would inform the king or a royal minister. . . . Then the king or the royal minister would renovate the city, and some time later that city would become successful and prosperous, well populated, filled with people, attained to growth and expansion. . . .
>
> So too, monks, I saw the ancient path, the ancient road traveled by the perfect buddhas of the past. And what is that ancient path, that ancient road? It is just this noble eightfold path: that is, right view, right intention, right speech, right action, right livelihood, right effort, right mindfulness, right concentration. . . . Having directly known them, I have explained them, to the monks, the nuns, the lay male disciples, and the lay female disciples. This holy life, monks, has become successful and prosperous, extended, popular, widespread, well proclaimed among gods and humans.[20]

Thus, in Buddhism, the truth is something that is found, and then lost, and then found again. This is why it is said that the next buddha does not appear in the world until the teachings of the previous buddha have been completely forgotten. As long as the path to the city of reality remains passable and the city itself remains prosperous, there is no reason for repair. But when the city falls into ruins and the path is overgrown with oblivion, then the path must be cleared again and the city restored. This is what the buddhas do, again and again, over the aeons.

Scholars have long held that the buddhas who preceded Śākyamuni (typically counted as three or six) were not historical figures; each is separated from his predecessor by billions of years. Instead, it is argued that the previous buddhas were introduced into the early tradition to counter charges that the Buddha was guilty of the crime of innovation. The Buddha's opponents in the Brahmanical traditions of ancient India claimed that their sacred texts, the Vedas, were eternal. They were never composed but had always existed, only having been heard by the ancient sages. The existence of the previous buddhas allows the Buddhists to make a similar claim. These four or seven buddhas (counting Śākyamuni) are the previous buddhas of our particular world system. Before that world system came into existence, there were previous worlds with their own buddhas, without beginning. The city in the forest was never built. It is an eternal city, to be discovered in each age.

The common portrayal of the Buddha in the literature of Buddhism and Science is quite different. Here the Buddha is seen as a scientist, experimenting in his laboratory of the spirit, trying first the life of indulgence and then the life of asceticism, testing the various meditative techniques of the day before sitting down under the tree on that full-moon night and discovering the truth for himself. The story is sometimes told in this way, and perhaps this is what indeed occurred at that unrecoverable moment, subject to such endless commentary. But whether he was the thousandth or the fifth or the first to see that truth, once it was seen, there has been no further truth to discover over the past two and a half millennia. To search for the truth in Buddhism is to follow the path to the ancient city, the path that the Buddha—whether discovered or uncovered—revealed to the world. There is no other path, there is no other city. In this sense, Buddhism is a profoundly conservative tradition, constantly decrying innovation as deviation from the path. Innovation has, of course, occurred in myriad ways over the course of the tradition, but that innovation must always be portrayed as elaboration, as yet another articulation of the Buddha's silent enlightenment. The content of this enlightenment is not regarded as a vague truth, the ever-receding end point of an endless path; the content of the Buddha's enlightenment is described in detail in the various Buddhist traditions. It is not, as the Victorians delighted in declaring, that Buddhism has no dogmas. It is perhaps that it has too many.

Yet there is a certain parallel between the Buddha of the tradition and the Buddha of science. The Buddha of the tradition is validated by being the last or, more accurately, most recent in a long line of enlightened beings who have discovered, and taught, the same truth. The Buddha of science is validated not by being at the end, but at the origin, as the perfected person who discovered truths that lesser men would only learn millennia later. For the Buddha of the tradition to be valid, he must have understood what others had known long before him. For the Buddha of science to be valid, he must have understood what others did not know, and would not know, until long after him.

Each of these visions is profoundly retrospective; each evinces a deep longing for the primordial. The authority of the Buddha of the tradition derives from the fact that he has simply rediscovered eternal truths that the prehistoric buddhas had also found; much of the early literature recounts their lives more than they do his. And the disciples of the Buddha of science derive deep comfort from the thought that modern discoveries in quantum physics were known by the ancient Buddha, so long ago.

Returning to the Dalai Lama, in the first chapter of *The Universe in a Single Atom*, he raises the fascinating question of the relation of experience and scripture in Buddhism.

> Although Buddhism has come to evolve as a religion with a characteristic body of scriptures and rituals, strictly speaking, in Buddhism scriptural authority cannot outweigh an understanding based on reason and experience. In fact, the Buddha himself, in a famous statement, undermines the scriptural authority of his own words when he exhorts his followers not to accept the validity of his teachings simply on the basis of reverence to him. Just as a seasoned goldsmith would test the purity of his gold through a meticulous process of examination, the Buddha advises that people should test the truth of what he has said through reasoned examination and personal experiment. Therefore, when it comes to validating the truth of a claim, Buddhism accords greatest authority to experience, with reason second and scripture last.[21]

There is much to consider here. Buddhism does indeed accord great authority to experience, that is, to the experience of the Buddha.[22] One of the most commonly cited qualities that distinguishes the enlightenment of the Buddha from the enlightenment of his disciples is that the Buddha came to his understanding of the nature of reality through his own efforts, without relying on the teachings of another buddha.[23] In the Buddha's previous life as the yogin Sumedha, he encountered a buddha of the past named Dīpaṃkara and realized that if he became his disciple, he could realize nirvāṇa and achieve liberation from rebirth in that very lifetime. Yet he decided to postpone his enlightenment by billions of years in order to become a buddha when there was no buddha in the world. Sumedha lay upon the ground and spread his matted locks across the mud so that Dīpaṃkara would not soil his feet. As he lay in the mud, he began to reflect (in Henry Clarke Warren's 1896 translation from the Pāli).

While thus I lay upon the ground,
Arose within me many thoughts:
"Today, if such were my desire,
I my corruptions might consume.

"But why thus in an unknown guise
Should I the Doctrine's fruit secure?
Omniscience first will I achieve,
And be a Buddha in the world.

"Or why should I, a valorous man,
The ocean seek to cross alone?
Omniscience first will I achieve,
And men and gods convey across."[24]

Thus, the bodhisattva waits to achieve enlightenment until a point in the distant future when there is no buddha to set forth the path. He discovers the path to the ancient city by himself. Others, however, must rely on the teachings of the Buddha to experience enlightenment, and even then, that experience is not said to be self-validating. The Buddha's foremost disciples, Śāriputra and Maudgalyāyana, who flank him in so many paintings and statues, had to be informed by the Buddha that they had reached the stage of arhat.

In the absence of the Buddha, one must rely on the sūtras. But the Buddha is said to have taught eighty-four thousand doctrines as antidotes to eighty-four thousand afflictions. And even if such a number is not immediately enumerated in the available texts, hundreds of discourses are counted as the word of the Buddha (*buddhavacana*) even by the Theravāda tradition, which rejects the large body of Mahāyāna sūtras as spurious. The Dalai Lama, however, is a proponent of the Mahāyāna, which vastly increases the size of the scriptural corpus. It is also acknowledged, in all forms of Buddhism, that the Buddha did not teach the same thing to each person he encountered, that he adapted his message to the interests and capacities of his audience. As a result, if every statement attributed to the Buddha is accepted literally, contradictions are immediately evident. The Buddha, however, as an enlightened being, must be free from all contradiction. Thus, interpretation is called for to resolve the apparent contradictions. The science of interpretation is highly developed in Buddhism, with various schools and thinkers across the Buddhist world and across the millennia declaring what the Buddha in fact had in mind when he set forth a particular doctrine.[25]

To understand the nature of knowledge and the role of experience in arriving at knowledge of the truth in Buddhism, it is perhaps useful to introduce one of the standard categories of Buddhist epistemology. This is the division of everything that can be known into three groups: the manifest (*abhimukī*, *mngon gyur*), the hidden (*parokṣa, lkog gyur*) or sometimes in Tibetan "slightly hidden" (*cung zad lkog gyur*), and the very hidden (*atyantaparokṣa*, *shin tu lkog gyur*). The category of the manifest includes those objects of knowledge that can be apprehended through direct perception, such as the accurate perception of the color or shape of an object by the eye consciousness. The second category, the hidden, includes those objects of knowledge, or facts, that cannot be perceived by direct perception but can be known through inference. The standard example is seeing smoke rising from a distant mountain pass and inferring the existence of a fire burning there. However, the category of the hidden includes several of the most fundamental doctrines of Buddhism, including the existence of liberation from rebirth, the possibility of omniscience (or buddhahood), the subtle impermanence of all conditioned things, and reincarnation (to be considered below). The Buddhist claim is that none of these can be seen directly by an unenlightened person, but each of these can be proved by reasoning. The third category, the very hidden, includes those things that remain inaccessible to the unenlightened through either direct perception or inference. These include the features of the various heavens located on and above Mount Meru as well as the subtle workings of the law of karma. Although the basic fact of karma—that virtuous deeds result in future happiness and nonvirtuous deeds result in future suffering—is merely hidden, and thus accessible to reasoning, the subtle workings of karma are very hidden. The Buddha is often portrayed explaining the particular deed that a given individual performed in a past life which resulted in a particular situation in the present life. The Tibetan textbooks on logic and epistemology note that the Buddha is able to explain in detail the particular deeds that produced each of the colors in the feathers of a peacock's tail.

Such information is available only to a buddha. All others must rely on what is called in Tibetan inference through belief (*yid ches rjes dpag*); the Sanskrit term translated as "belief" is *āpta*, which denotes that which is authoritative, credible, trustworthy, believable. Inference through belief is gained not through experience, or reasoning, but through reliance on scripture; it is also called inference based on scripture (*āgamāśritānumāna*, *lung gi rjes dpag*). Here, then, knowledge is to be gained through what is termed an "incontrovertible scripture," which has three qualities, known as the three analyses (*dpyad pa gsum*): (1) in its statements about manifest phenomena, it is not contradicted by direct perception; (2) in its statements about hidden phenomena, it is not contradicted by inference based on reasoning or "the power of the fact"; and (3) in its statements about the very hidden, those things inaccessible to ordinary direct perception and inference, it is not contradicted by other incontrovertible scriptures. Despite the apparent circularity of such an analysis, the important point is the central role that the authority of scripture plays in Buddhist thought.[26]

The categories of the manifest, the hidden, and the very hidden provide the context when the Dalai Lama notes that:

> From the Buddhist point of view, there is a further level of reality, which may remain obscure to the unenlightened mind. Traditionally, a typical illustration of this would be the most subtle workings of the law of karma, and the question of why there are so many species of beings in the world. Only in this category of propositions is scripture cited as a potentially correct source of authority, on the specific basis that for Buddhists, the testimony of the Buddha has proven to be reliable in the examination of the nature of existence and the path to liberation.[27]

Despite its understatement, this is a crucial point. By the "most subtle workings of the law of karma," the Dalai Lama refers not to the law of karma in general, the "natural law" according to which happiness is the result of virtuous deeds and suffering is the result of evil deeds; according to Buddhist logicians, the existence of the law of karma can be inferred through proper reasoning. He refers instead to the "very hidden," that is, those things that are inaccessible to the direct perception and inference of the unenlightened, and are known only to a buddha. To understand such things, as noted above, one must rely on the statements of the Buddha, which are recorded in the scriptures.

Perhaps a modern correlate to the colors in a peacock's tail would be the DNA sequencing of a genome. The Buddhist answer to why there are so many different species in the world—a question answered by science with the theory of natural selection—would also be the law of karma; the physical forms of the beings in the universe are the direct results of deeds done in the past. As a famous Buddhist saying succinctly explains, "If you wish to know what you did in the past, look at your present body. If you wish to know what you will look like in the future, look at your present mind." In a certain sense, then, the questions that can only be answered by consulting the Buddhist scriptures are not only central to Buddhist doctrine; they are also central to science. As the Dalai Lama notes, Buddhist philosophers have argued that the Buddha may be confidently relied upon concerning these very subtle questions because he has been shown to be reliable concerning the most central questions: the nature of reality and the path to liberation from suffering and rebirth. In a widely quoted passage, the Madhyamaka master Āryadeva declares (at *Catuḥśataka* 12.5), "Whoever has doubts about the Buddha's statements concerning the hidden, will trust only him based on his [teaching of] emptiness."[28]

Thus, the centrality of scripture in Buddhism is difficult to overstate. Significantly, when discussing the importance of experience over scriptural authority in Buddhism, the Dalai Lama does not describe his own experience; to do so would be deemed inappropriate in a public context. Instead, he cites a Buddhist scripture that instructs monks on how to regard the scriptures of the Buddha. It is a commonly quoted passage, which reads: "O monks, like gold that is heated, cut, and rubbed, my words should be analyzed by the wise and then accepted; they should not do so out of reverence."[29] This passage, at least to my knowledge, has not been located in a sūtra, that is, in a text that the tradition ascribes to the Buddha. It appears instead in a famous treatise (śāstra) of late Indian Mahāyāna, the *Tattvasaṃgraha* (Compendium of Principles) by the eighth-century Bengali master Śāntarakṣita, known in Tibet as the "bodhisattva abbot" for his efforts in founding the first Buddhist monastery there. A massive work of 3,646 verses in 26 chapters, it is a polemical survey of the philosophical positions held by a wide variety of non-Buddhist (and some Buddhist) schools on a number of topics or principles (*tattva*), in which Śāntarakṣita demonstrates their faults.

The famous passage about gold occurs near the end of the text. Śāntarakṣita is arguing that the Vedas are not a valid source of knowledge, but that the word of the Buddha is. In order to claim this, he must establish the possibility of a person achieving omniscience. He begins with the logical point that the mere fact that his Hindu opponents have never perceived an omniscient person does not establish the impossibility of such a person existing; indeed, the omniscient person can only be apprehended by another omniscient person.[30] He goes on to argue that an omniscient person is to be judged by his or her knowledge of the truth. Thus, Śāntarakṣita asserts that the Buddha is omniscient not because of who he was but because of what he taught: *anātman*, the doctrine of no-self, a doctrine unique among all teachings.[31] He goes on to provide a fairly standard Mahāyāna list of the qualities of the Buddha and his extraordinary pedagogical skills (such as those outlined in chapter 1): that he teaches the dharma without the slightest operation of thought, like a wheel set in motion;[32] that he is not subject to the faults of mortal beings because he is beyond the cycle of rebirth and thus immortal.[33]

The specific context of the passage is this: Śāntarakṣitsa is mocking the Hindu scriptures, including the Vedas, and the teachers who taught them. These teachers, he says, suspected that the Vedas were false, but in order to hide this fault they taught them only to the brahman caste, because its members were particularly dull-witted. They furthermore declared that these texts were commandments to be accepted without question and never subjected to reasoning. The teachings of the buddhas, however, are utterly different.

> Whatever is spoken by the great beings is endowed with reason; they are certain in their own ability to elucidate it. Without the slightest fear, the roar of the lion withers the arrogance of the evil *tīrthika*s, drunken elephants, in this way: "O monks, like gold that is heated, cut, and rubbed, my words should be analyzed by the wise and then accepted; they should not do so out of reverence."[34]

Thus, unlike the scriptures propounded by Hindu teachers, the teachings of the buddhas are endowed with reason (*yuktārthatvaṃ*); hence the buddhas have no fear in setting them forth. Śāntarakṣita compares the Hindu teachers, called *tīrthika*s, to drunken elephants, intoxicated into falsely believing in their own authority. But their arrogance is destroyed when they hear the lion's roar, a common metaphor for the Buddha's teaching. Unlike the Hindu teachers, the Buddha does not fear that his words will be subjected to analysis. Indeed, he encourages his monks to do so before accepting their authority. The lion's roar, therefore, is not an instruction to analyze the words of the Buddha and then accept them or reject them. It is rather the Buddha's boast, as portrayed by Śāntarakṣita: confident in the authority of his teachings, the Buddha challenges his monks to analyze his words before accepting them. He has no fear that they will analyze his words and then somehow decide not to accept them. After heating it, slicing it, and polishing it, the goldsmith will find that the shiny yellow metal is in fact gold.

Śāntarakṣita goes on to explain that the scriptures attributed to the Buddha need not have been actually spoken by him; sometimes they even emanate from walls. Hence, the Buddha is not to be regarded as the author of the sūtras; they are rather set forth under his supervision.[35] Reiterating the Buddhist view of the omniscience of the Buddha described in chapter 1, Śāntarakṣita asserts that the Buddha comprehends everything that exists in a single instant, without needing to know them sequentially, unless that is his wish.[36]

The Dalai Lama explains that a fundamental attitude of Buddhism is the "commitment to keep searching for reality by empirical means and to be willing to discard accepted or long-held positions if our search finds the truth is different."[37] As the preceding discussion suggests, these empirical means are provided in the Buddhist scriptures. This might be further illustrated by examining the category of the three types of wisdom: the wisdom arising from hearing, the wisdom arising from thinking, and the wisdom arising from meditation.

The wisdom arising from hearing refers to the level of understanding that one can gain through study; *hearing* in this context refers specifically to listening to Buddhist teachings, but also extends to include the reading of Buddhist texts. The second form of wisdom arises from thinking. Here *thinking* refers to what one would normally call meditation: the understanding that results from a careful and systematic investigation of what has been studied, carried out while seated in the formal meditative posture. The third and highest form of wisdom is the wisdom arising from meditation, where *meditation* refers specifically to understanding conjoined with the deep level of concentration known as serenity (*śamatha*). Using this concentrated mind to understand the absence of self produces a state called insight (*vipaśyanā*), where the long-held conviction that the self exists can be discarded. It is only when this insight is transformed into a state called yogic direct perception that the seeds of future suffering can be permanently destroyed. *Hearing*, a term that reflects the Indian emphasis on sound (shared by both Hinduism and Buddhism), means listening to the teachings of the Buddha; the early disciples of the Buddha were called *śrāvaka*s, "listeners." The words of the Buddha were committed to writing to produce the Buddhist scriptures. It is the study of these scriptures, rather than experience, that results eventually in the "wisdom arisen from hearing," the necessary prerequisite for higher meditative states. And one of the central tenets of Buddhism that must be heard is the doctrine of karma.

An attraction of Buddhism to European intellectuals during the Victorian period was that it presented an ethical system that did not require God, yet somehow seemed consistent with Darwinism. As the leading British scholar of Buddhism of the day, Thomas W. Rhys Davids, stated in his Hibbert Lectures of 1881, "And the more thorough-going the Evolutionist, the more clear his vision of the long perspective of history, the greater will be his appreciation of the strangeness of the fact that a theory so far consistent with what he holds to be true should have been possible at all in so remote a past."[38] Karma took the place of divine reward and retribution, and in the discourse of Buddhism and Science, karma would be described as a "natural law." Thomas Huxley wrote in 1894: "Like the doctrine of evolution itself, that of transmigration has its roots in the world of reality; and it may claim such support as the great argument from analogy is capable of supplying.[39]

Declarations of similarity between the laws of karma and rebirth on the one hand and the law of natural selection on the other have been a commonplace since the Victorian period; each seems to be able to account for the creation of life-forms without recourse to a creator. But how compatible are karma and evolution? One of the few Buddhist voices to raise this question is that of the Dalai Lama: "From the Buddhist perspective, the idea of these mutations being purely random events is deeply unsatisfying for a theory that purports to explain the origin of life."[40]

The doctrine of karma predates the appearance of the Buddha in India. The term, which simply means "action" in Sanskrit, referred especially to the performance of rituals. Ritual deeds, such as the fire sacrifice, were efficacious when done well, while ritual deeds that were somehow mishandled did not produce the desired result. The Buddha is sometimes credited with moving karma from the physical and ritual realm into the mental and ethical realm (although such views of karma are also found in the Upaniṣads). It is clear, however, that for the Buddha the notion of intention was central to his understanding of karma, and that karma was central to his teaching.

The most famous statement in Buddhism is not the *Heart Sūtra*'s declaration that "form is emptiness, emptiness is form." This is a statement from a Mahāyāna sūtra, and thus, although it is chanted daily in Buddhist monasteries in China, Japan, Korea, and Tibet, it is not accepted as the word of the Buddha by the Theravāda traditions of Sri Lanka and Southeast Asia. Perhaps the most famous statement in Buddhism is one found inscribed and recited across Buddhist Asia. It is the somewhat unlikely declaration, "For those things that have causes, he has shown their causes. He has also shown their cessation. Thus, has the great sage spoken." According to the traditional account, this is how one of the first disciples summarized the Buddha's teaching to a passerby. The person to whom this statement was addressed, Śāriputra, achieved the first level of enlightenment simply upon hearing these words. The statement is found throughout the Buddhist world, from Pāli chants to tantric sādhanas, and was sometimes inscribed on paper and placed inside stūpas as a substitute for a relic of the Buddha. It suggests that, more than the doctrine of no-self or the doctrine of nirvāṇa (which the statement implies with its reference to "cessation"), what was deemed most important about the Buddha's teaching, at least in the early tradition, was his emphasis on causation, and that by identifying the cause of a particular effect and then eliminating that cause, the effect can be forever prevented. This is seen most clearly in the doctrine of the four truths.

The first truth declares that existence in the realm of rebirth is qualified by suffering. Various forms and levels of suffering are enumerated. One of the most important of these is three types of suffering: the suffering of pain, the suffering of change, and the suffering of conditioning. The suffering of pain comprises physical and mental feelings of pain. The suffering of change comprises physical and mental feelings of pleasure; the Buddhist claim is that all feelings of worldly pleasure will eventually turn into pain. The suffering of conditioning is the most subtle form of suffering: it is the fact that beings in the cycle of rebirth are so conditioned as to be susceptible to suffering in the next instant.

All suffering is the result of negative deeds and the negative mental states (*kleśa*) that motivate them. These constitute the second of the four truths, the truth of origin. Physical, verbal, and mental deeds motivated by negative emotions such as desire, hatred, and ignorance produce suffering in the future, both in the present life and in future lives. Indeed, all feelings of physical and mental pain are the result of deeds performed in the past. Thus, suffering is produced by negative deeds, negative deeds are motivated by the negative mental states of desire and hatred, and desire and hatred are produced by ignorance, the belief in a permanent and autonomous self. If ignorance can be destroyed, desire and hatred become impossible. Without desire and hatred, no negative deeds are performed. Without negative deeds, there is no karma to produce future suffering. "For those things that have causes, he has shown their causes. He has also shown their cessation." That cessation of suffering is the third truth, the truth of cessation. The fourth truth is the path to that cessation.

In Buddhism, therefore, karma is the engine of the cycle of existence. According to the classical formulations of the Abhidharma, the past deeds of the beings reborn there create their experiences of pleasure and pain, the bodies and minds that undergo those experiences, and the physical domains that those beings inhabit. Without karma, the cycle of existence would cease to exist. And its cessation, called nirvāṇa, is the ultimate goal of Buddhist practice. The precise relation between the path and the goal preoccupied Buddhist thinkers across Asia.

The cycle of rebirth called *saṃsāra* and the myriad sufferings that occur there are the raison d'être for the Buddha's teaching. It was what motivated the prince to leave the palace, to practice austerities for six years, to achieve enlightenment, and to teach the dharma. Suffering and rebirth are the products of karma. If the forms of life on earth, the animal and human species, are instead shown to be the result of evolution, of a process of natural selection, karma becomes superfluous and rebirth becomes impossible.

Darwin's theory of evolution thus presents particular problems for Buddhism because it obviates the law of karma. This is why the Dalai Lama raises questions about natural selection and what he calls "the emergence of human consciousness." As he explains in *The Universe in a Single Atom*, the classic proof of rebirth is provided by the seventh-century philosopher Dharmakīrti. In his most famous work, the *Pramāṇavārttika* (Commentary on Valid Knowledge), he argues that cause and effect must be of the same substance; a seed is the same substance as the sprout it produces, a pot is the same substance as the clay from which it is produced. It is a fundamental tenet of Buddhist philosophy that mind and matter, although closely related in many ways, are substantially different, such that mind cannot be produced from matter. As the Dalai Lama states, "According to Buddhism, though consciousness and matter can and do contribute toward the origination of each other, one can never become the substantial cause of the other."[41] Consciousness is the product of consciousness. The present moment of consciousness is the product of a previous moment of consciousness. The consciousness of a newborn infant is the product of the consciousness of the infant in the womb. The consciousness in the womb is the product of the consciousness present at the moment of conception. The consciousness present at the moment of conception is the product of the consciousness at the moment prior to conception. That moment of consciousness can only come from a previous lifetime. Hence, according to Dharmakīrti, rebirth has been proved. Furthermore, in Buddhism rebirth, and hence consciousness, has no beginning.

As the Dalai Lama notes, the Buddha is said to have refused to answer questions about origins and end points; such concerns are counted among the fourteen questions, called the "unindicated views" (*avyākṛta*), to which the Buddha remained silent. His silence is widely interpreted: some say that he did not answer because such questions are irrelevant to the immediate task of liberation from rebirth; some say that he did not answer because an answer of yes or no could be misconstrued; some say that the questions themselves had ontological implications that the Buddha did not accept. This places the Buddhist thinker in a difficult position regarding neurobiological investigations of the origin and nature of consciousness. The Dalai Lama writes, "But assuming mind is reducible to matter leaves a huge explanatory gap. How do we explain the emergence of consciousness? What marks the transition from nonsentient to sentient beings? A model of increasing complexity based on evolution through natural selection is simply a descriptive hypothesis, a kind of euphemism for 'mystery,' and not a satisfactory explanation."[42] In Buddhism, consciousness does not emerge; it has no beginning. There is no transition from nonsentient to sentient beings; since mind and matter are substantially different, the nonsentient cannot become sentient.

This may be one reason why the Dalai Lama expresses such caution on the issue of cloning, which wreaks havoc upon Buddhist doctrines of karma and rebirth. Perhaps the closest analog to cloning in Buddhist doctrine is the magical power of emanation, by which buddhas and highly advanced bodhisattvas create multiple forms of themselves in order to serve suffering sentient beings. But these multiple forms all remain aspects of the same person. And although buddhas can appear in whatever form is appropriate to benefit the world—animate or inanimate—it is difficult to imagine a situation in which the Buddha would appear as a sheep.

Buddhist doctrine rejects the existence of a permanent, partless, independent self, but maintains the category of the person. That is, each sentient being is an individual stream, a combination of mind and matter, accumulating karma and experiencing its fruits over the course of billions of lifetimes, until (according to some schools) each of those streams of mind and matter becomes a buddha. With this view of the person, what would occur when an animal was cloned?

According to the law of karma, experiences are the result of past deeds. Would a cloned sentient being carry the same karma as the original sentient being? Would they have identical experiences? If so, two sheep, the original and the clone, should feel frisky or sleepy at the same time, should get hungry at the same time, should get shorn at the same time, should give birth to identical lambs at the same time, should go to the slaughter at the same time. But the clone had to grow from a lamb in order to become identical in form to her original. Does this mean that there would be a time lag in these experiences? One might also speculate about what the first sheep did in a past life that resulted in its being cloned. Was it a good deed or a bad deed? Perhaps these are the kinds of questions that the Buddha said "tend not to edification."

The Dalai Lama speaks of the "limits of scientific knowledge" regarding questions of the nature of consciousness, questions that are central to the Buddhist doctrines of karma and rebirth; for him rebirth is something that science has not found, but also has not found to be nonexistent. He is critical of the theory of natural selection on two counts. First, the random nature of mutations is at odds with the doctrine of karma, which he defends at several points: "From the scientific view, the theory of karma may be a metaphysical assumption—but it is no more so than the assumption that all of life is material and originated out of pure chance. . . . I believe that karma can have a central role in understanding the origination of what Buddhism calls 'sentience,' through the media of energy and consciousness."[43]

Second, the Dalai Lama finds fault with the theory of natural selection because the competition for reproduction it postulates seems to preclude the possibility of altruism; he sees a danger in human beings being "reduced to nothing more than biological machines, the products of pure chance in the random combination of genes, with no purpose other than the biological imperative of reproduction."[44] Without karma, there can be no rebirth. Without rebirth, there is no realm of saṃsāra. Without saṃsāra, there is no compassion, or at least the extraordinary compassion of the bodhisattva, who vows to liberate all beings in the universe from the sufferings of the realm of rebirth. It was compassion that motivated the Buddha to perfect himself over millions of lifetimes. It is compassion that defines the Dalai Lama as the incarnation of Avalokiteśvara, the bodhisattva of compassion, who enters a human womb in order to be reborn among humans, not by chance but by choice.

Thus, the Dalai Lama speaks repeatedly of a "fruitful collaboration" between Buddhism and Science. He is willing to concede the scientific view on almost all questions of cosmology. At the same time, he holds firmly to the doctrine of karma and the immaterial nature of consciousness. He clearly feels that there are questions crucial to scientific exploration to which Buddhism can contribute. Indeed, he seems to anticipate a kind of "paradigm shift," one that will confirm what he sees as the fundamental truths of Buddhism. In Mahāyāna Buddhism, the dyad of wisdom and compassion is consistently described as central to the achievement of enlightenment. The Dalai Lama seems to describe a new Buddhism, one that retains compassion as it primary motivation, but adds the discoveries of modern science to the wisdom needed to complete the long path to buddhahood—something presumably not needed by premodern aspirants to the state of unsurpassed perfect enlightenment.

It is as if having successfully won the battle against the Christian missionaries a century ago, Buddhism must now accommodate the findings of modern science if it is not to be dismissed as an anachronism. As Gendun Chopel wrote in 1941, "Having become an open-minded person who sees the important and unimportant, you should strive to ensure the survival of the teaching [Buddhism], so that it remains together with the ways of the new reasoning. . . . Please pray that the two, this modern reasoning of science and the ancient teachings of the Buddha, abide together for tens of thousands of years."

This is a prayer for a new Buddhism. But what is to become of the old Buddhism? Nowhere in *The Universe in a Single Atom* does the term *nirvāṇa* appear. Rebirth is mentioned, but not the six places of rebirth: as gods, demigods, humans, animals, ghosts, and hell beings. Where are the deities who animate the landscape and the divine protectors whom the Dalai Lama consults when making a momentous decision? Where is the Buddha's relentless disparagement of the cycle of rebirth, a world in which beings are so conditioned as to be susceptible to suffering in the next instant, a world with sufferings so vast and deep that when describing them the Dalai Lama will sometimes cover his head with his monk's robe and weep? Where is the uncompromising assertion that this world is built by ignorance, a world that ultimately is not to be improved, but from which one must seek to escape, along with all other beings, with the urgency that a person whose hair is ablaze seeks to douse the flames? Where is the insistence that meditation is not intended to induce relaxation but rather a vital transformation of one's vision of reality? Is this Buddhism placed at risk by the compulsion to find convergences with Science, this Buddhism that makes the radical claim that it is possible to live in the world untainted by what are called the eight worldly concerns: gain and loss, fame and disgrace, praise and blame, happiness and sorrow? These are the teachings that the greatest Buddhist thinkers, figures such as Nāgārjuna and Dharmakīrti, accepted implicitly. These are the teachings that the Dalai Lama has offered with unrivaled eloquence for so long.

THE SCIENCE OF BUDDHISM

Harold Fielding-Hall (1859–1917) was a British officer in the Third Anglo-Burmese War, during which Burma was annexed by Great Britain. After the war, he remained in Burma as a local district magistrate, serving from 1887 to 1891. Upon his return to Britain, Fielding-Hall published *The Soul of a People*, a book about Burmese Buddhism, where he writes:

> At first sight it seems that of all creeds none is so full of miracle, so teeming with the supernatural, as Buddhism, which is, indeed, the very reverse of the truth. For to the supernatural Buddhism owes nothing at all. It is in its very essence opposed to all that goes beyond what we see of earthly laws, and miracle is never used as evidence of the truth of any dogma or of any doctrine.
>
> If every supernatural occurrence were wiped clean out of the chronicles of faith, Buddhism would, even to the least understanding of its followers, remain exactly where it is. Not in one jot or tittle would it suffer in the authority of its teaching.[1]

Despite living for years in a country where Buddhist practice, Buddhist monks, and Buddhist institutions played such a central role,

Fielding-Hall could claim that in Buddhism there was no prayer, no ceremonial, "no praise, no thanksgiving of any kind."[2] This could have occurred only because, by the time that Fielding-Hall published his book, the Buddha had been transformed from stone idol into a man of flesh and blood, an earnest seeker of the truth. In a sense, the stone Buddha of India, where Buddhism was long extinct, had been brought to unnatural life by a strange process of scientific reanimation; the dead was made living by a new science. This new Buddha, the Buddha of Fielding-Hall and of his fellow Victorians, was vivified by European scholarship, the academic study of Buddhism or, as it was called in those days, "the scientific study of Buddhism." It was this Western science, fueled by the study of dead languages, that built a Buddha whose teachings could be compatible with science. Indeed, Fielding-Hall hails him as "this Newton of the spiritual world."[3]

The story of the European encounter with, and adoption of, the Buddha is too long to tell here. However, to understand the process of Buddhism's confluence with Science, it is important to have some sense of how the Buddha of that Buddhism came to be. The story will begin in Oxford in the late nineteenth century and then go back in time to Kathmandu and Paris in the first half of that century, before returning to England and a politely rancorous exchange between two leading figures in the Victorian representation of Buddhism, published in the pages of a periodical appropriately named *The Nineteenth Century*.

In 1888 two of the most significant figures in the modern history of Buddhism met in Oxford. The first was Friedrich Max Müller (1823–1900), professor of comparative philology and the most famous orientalist of the nineteenth century. Müller was a German-born Sanskritist who spent most of his life in England, devoting his considerable talents to what he regarded as his life's work, and for which he is perhaps least remembered: the production of a critical edition of the Rig Veda, a task that had been assigned to him by his Sanskrit teacher in Paris, Eugène Burnouf. In 1860 Müller was denied the Boden Professorship in Sanskrit, presumably because he was neither English nor Anglican (and this was the only sense in which he was a Nonconformist), in favor of Monier-Williams (1819–1899), his linguistic inferior, whose Sanskrit studies were motivated largely by his commitment to the conversion of India to Christianity. But another professorship was created for Müller, and from it he produced a prodigious scholarly deposit, including, most notably for the public, the fifty-volume Sacred Books of the East series in 1894.

Ten of its forty-nine volumes were devoted to Buddhist works. Reflecting the opinion of the day that Pāli texts of the Theravāda tradition represented the most accurate record of what the Buddha taught, seven of these volumes were devoted to Pāli works. Among other Indian texts, Aśvaghoṣa's famous life of the Buddha appeared twice, translated in one volume from Sanskrit and in another from Chinese. The *Lotus Sūtra* was included in another volume. The final volume of the series, entitled *Buddhist Mahāyāna Texts*, contains such famous works as the *Diamond Sūtra*, the *Heart Sūtra*, and the three Pure Land sūtras. The presence of this array of Buddhist texts in Müller's series attested to the philological skills developed by European orientalists over the course of the nineteenth century. At his home in Oxford in 1888, Müller entertained the American Theosophist Colonel Henry Steel Olcott (1832–1907).

Although already described in the introduction, let me briefly reiterate something of Olcott's life and works here. Raised in a Presbyterian family in New Jersey, Olcott developed an interest in spiritualism at an early age. He served in the Union army during the American Civil War and subsequently was appointed to the commission that investigated the assassination of Lincoln. Working as a journalist in New York City, he occasionally reported on "spiritualism," the beliefs and practices connected with communicating with the spirits of the dead—something very much in vogue in the last half of the nineteenth century. In 1874 he made a trip to Chittenden, Vermont, to investigate paranormal events occurring in a farmhouse belonging to the Eddy brothers, who were said to be able to summon several spirits, including that of an Indian chief named Santum. There he met the Russian émigré and medium Helena Petrovna Blavatsky (1831–1891). Their shared interest in spiritualism, psychic phenomena, and esoteric wisdom led them in 1875 to found in New York the Theosophical Society, an organization that would bring the teachings of the Buddha, at least as interpreted by the society, to a large audience in Europe and America over the subsequent decades. For Blavatsky and Olcott, Theosophy was an ancient wisdom that was the root and foundation of the world's mystical traditions. This wisdom had been dispensed over the millennia by a group of Atlantean masters called mahatmas, or "great souls." In the modern period, these masters had congregated in a secret location in Tibet. Madame Blavatsky claimed to have studied under their tutelage over the course of seven years there and to have remained in psychic communication with them.

Having corresponded with the Hindu reformer and founder of the Arya Samaj, Swami Dayanda Saraswati (1824–1883), Blavatsky and Olcott sailed to India, arriving in Bombay in 1879. Wishing also to establish ties with Buddhist leaders, they proceeded the next year to Ceylon, where they took the vows of a lay Buddhist; Olcott was presumably the first American to do so. He enthusiastically embraced his new faith, which he felt contained no dogma that he was compelled to accept. Shocked at what he perceived as the ignorance of the Sinhalese about their own religion, Olcott took it as his task to restore true Buddhism to Ceylon and to counter the efforts of the Christian missionaries on the island. In order to accomplish this aim, he adopted many of the missionaries' techniques, founding lay and monastic branches of the Buddhist Theosophical Society to disseminate Buddhist knowledge (and later assisting in the founding of the Young Men's Buddhist Association). In 1881 he published *A Buddhist Catechism*, a series of questions and answers about Buddhism. The work was translated into Sinhalese and memorized by Sri Lankan children.

In 1885 Olcott set out on a mission to heal the schism he perceived between "the Northern and Southern Churches," that is, between the Buddhists of Ceylon and Burma (Southern) and those of China and Japan (Northern). He believed that a great rift had occurred in Buddhism twenty-three hundred years earlier, and that if he could simply have representatives of the Buddhist nations agree to a list of shared doctrines, it might be possible to create a "United Buddhist World." He was unsuccessful in the first attempt, but set out again in 1891, armed with a list of "fourteen items of belief" (he also referred to them as "Fundamental Buddhistic Beliefs"). He traveled to Burma, Sri Lanka, and Japan, where he negotiated with Buddhist leaders until he could find terms to which they could assent. The third of the fourteen beliefs, according to an 1894 version, is "The truths upon which the Dharma is founded are scientific. They have, we believe, been taught in successive ages (*kalpa*s), or prehistoric epochs, by certain fully illuminated beings defined as human Buddhas (*manushi-buddha*)."[4] Olcott's conviction that there was a single *Buddhism* to which all Buddhists of Asia could attest would come to underlie much of the Buddhism and Science discourse. Indeed, it is important to note that one of the first forms of *Science* with which Buddhism was said to be compatible was the science of Theosophy.

When Olcott visited Müller at his home in Oxford in 1888, Olcott was in the midst of his campaign for a United Buddhist World, while seeking also to mediate persistent squabbles within the Thesophical Society. He describes their meeting in his diary:

> Professor Müller was so kind as to say that the Oriental reprinting, translation, and publishing portion of the Society's work was "noble, and there could be no two opinions about it, nor were there among Orientalists." But as for our more cherished activities, the discovery and spread of ancient views on the existence of Siddhas and of the *siddhis* in man, he was utterly incredulous. "We know all about Sanskrit and Sanskrit literature," he said, "and have found no evidence anywhere of the pretended esoteric meaning which your Theosophists profess to have discovered in the Vedas, the Upanishads, and other Indian scriptures: there is nothing of the kind, I assure you. Why will you sacrifice all the good opinion which scholars have of your legitimate work for Sanskrit revival to pander to the superstitious belief of the Hindus in such follies?" We sat alone in his fine library room, well lighted by windows looking out on one of those emerald, velvety lawns so peculiar in moist England; the walls of the chamber covered with bookcases filled with the best works of ancient and modern writers, two marble statuettes of the Buddha sitting in meditation, placed to the right and left of the fireplace, but *on the hearth* (Buddhists take note) [.] . . . I see this greatest pupil of that pioneer genius, E. Burnouf, sitting there and giving me his authoritative advice to turn from the evil course of Theosophy into the hard and rocky path of official scholarship, and be happy to lie down in a thistle-bed prepared by Orientalists for their common use. . . . The Professor, finding me so self-opinionated and indisposed to desert my true colors, said we had better change the subject. We did, but not for long, for we came back to it, and we finally agreed to disagree, parting in all courtesy, and, on my part, with regret that so great a mind could not have taken in that splendid teaching of the Sages about man and his powers, which is of all in the world the most satisfying to the reason and most consoling to the heart.[5]

In his letter of thanks, Olcott asked that Müller consider moving the Buddha images that sat on the hearth in his study to a more exalted position, explaining that "the Buddhists are very sensitive about such things, and a painful impression would be made upon the mind of any sincere person of that faith, if he should call at your house and see them in your fireplace." Müller's wife reports that her husband "endeavoured to comfort Col. Olcott, by assuring him that with the Greeks the hearth was the most sacred spot, and this had induced him to place these Buddhas, which had been taken from the great Temple of Rangoon, in that position."[6]

This is a telling exchange. Although both were leading expositors of Buddhism, Max Müller and Henry Steel Olcott occupied different positions and inhabited different worlds: Müller, German expatriate and Oxford don, distinguished Sanskrit scholar (and student of the great Burnouf) who read Buddhist manuscripts in the original Sanskrit and Pāli, remembered today as the father of the "Science of Religion"; Olcott, American expatriate, committed Theosophist with no formal training in the classical languages of Buddhism. Olcott had traveled extensively in the Buddhist world and met with many leading monks; he is remembered today as the founder of a Victorian "spiritual science." Müller never traveled beyond Europe, and Buddhists were unlikely to call at his house; among the few Buddhists he encountered were Japanese scholars of the Pure Land sect who came to Oxford to study Sanskrit. Müller was the scholar par excellence, concerned with the historical reconstruction of the original teachings of Hinduism and Buddhism. Olcott was the devotee and enthusiast, less concerned with the form or even the surface content of the ancient texts, seeing them instead as repositories, when their symbolism was decoded, of the esoteric wisdom of Theosophy. Müller sought to dispel Olcott's irrational fantasies. Olcott lamented that so learned a scholar as Müller could not see the deeper meaning hidden on the page.

Olcott was not uncritical of the Buddhism he encountered in Asia. He came into conflict with some of the leading monks of Sri Lanka over what he considered the superstitious practice of worshipping the Buddha's tooth enshrined at Kandy; he claimed that it was not even a human tooth but a piece of deer horn. But he was not insensitive to Buddhist mores. Olcott knew from his years in Sri Lanka and his travels elsewhere in Asia that it was deeply offensive to allow anything associated with the dharma to touch the floor; one would not place a sūtra on the floor, for example, and one would never place a statue of the Buddha on the floor.

As a representation of an exalted being, the image of the Buddha must also be exalted. Despite his protestations to the Sinhalese about the excessive ritualism of their Buddhism, he nonetheless possessed a cultural sensibility, according to which an image of the Buddha was to be treated with the respect that a Buddhist would accord to it, whether that image was in a temple in Rangoon or a private home in Oxford.

In his response, Müller casually notes that the statues of the Buddha in his hearth indeed came from "the great temple of Rangoon" (presumably the Shwedagon). It is unclear whether they were pillaged during the First Anglo-Burmese War, when British troops captured and held the temple for two years, or during the Second, when British troops captured the temple in 1852; it then remained under the control of the military until 1929. Such was the confidence of the British Empire that Müller was not reluctant to acknowledge, tacitly, that the statues had been stolen from a Buddhist temple.

The more interesting element of Müller's response is that he had placed the statues of the Buddha on the floor because "with the Greeks the hearth was the most sacred spot." The comment sounds slightly disingenuous, but its implication is important. For Müller, the Buddha, removed from Asia and transported to England, is not Asian and therefore need not be bound by Asian custom. The Buddha is a figure of European culture, like a Greek god; and as the newest member of this ancient pantheon from which Western civilization emerged, he should be worshipped accordingly.

The Buddha's place in this pantheon is crucial to the discourse of Buddhism and Science. How, then, did he get there? Müller was a sometime participant and active eyewitness in this process of the Buddha's ascension. But the story begins long before his time.

In 1290 Marco Polo was making his way back to Venice after his years at the court of the great Kublai Khan. On the voyage west, his ship stopped in Sri Lanka, where he was told of a mountain at whose summit was a tomb. He writes:

> And I tell you that on this mountain is the sepulchre of Adam our first parent; at least that is what the Saracens say. But the Idolaters say that it is the sepulchre of Sagamoni Borcan, before whose time there were no idols. They hold him to be the best of men, a great saint, in fact, according to their fashion, and he was the first in whose name idols were made.[7]

Marco Polo is describing what is known as Adam's Peak, where there is no sepulcher but there are footprints, which the Muslims indeed claim to be those of Adam, and the Buddhists claim to be those of Śākyamuni Buddha, to whom Marco Polo refers by the Mongol version of his name. His account goes on to describe the famous story of Prince Siddhārtha's chariot rides that led him to renounce the world and set out on the path to enlightenment. He concludes, "And there he did abide, leading a life of great hardship and sanctity, and keeping great abstinence, just as if he had been a Christian. Indeed, and he had but been so, he would have been a great saint of Our Lord Jesus Christ, so good and pure was the life he led."[8]

Marco Polo's account was just one of many descriptions of the Buddha made by various European adventurers, diplomats, and missionaries prior to the nineteenth century, marked by varying degrees of what we would today judge as accuracy. He is usually described as an idol, and he is called by many names: Sagamoni Borcan, Xaca, Sommona Codom, Fo, Khodom, Boodhoo. It is only in the early eighteenth century that the conclusion began to be widely drawn that these names somehow referred to the same god, and that he may have been a historical figure. We read in volume 15 of Diderot and d'Alembert's *Encyclopédie, ou Dictionnaire raisonné des sciences, des artes et des métiers*, published in 1765, this entry on Siaka (that is, Śākya[muni], the "sage of the Śākya clan," one of the epithets of the Buddha). It is categorized under "Modern History of Superstition."

> SIAKA, religion of, (Hist. mod. Superstition) this religion, which is established in Japan, has as its founder *Siaka* or *Xaca*, who is also called *Budso*, & his religion *Budsodoism*. It is believed that the *buds* or the *siaka* of the Japanese is the same as the *foë* of the Chinese, & the *visnou*, the *buda* or *putza* of the Indians, the *sommonacodum* of the Siamese; for it seems certain that this religion came originally from the Indies to Japan, where they previously only professed the religion of the *sintos*.[9]

A century later, much had changed. In 1862 Max Müller published a review of Barthélemy Saint-Hilaire's *Le Bouddha et sa religion*. It was a long review, over fifty pages in length, and Müller used it as an occasion to survey the development of Buddhist studies in the previous decades of the nineteenth century. In the course of his survey, he describes the Sanskrit manuscripts that Brian Houghton Hodgson (1800–1894), British Resident at the Court of Nepal, had discovered in Kathmandu and then dispatched to Calcutta, London, and Paris. Müller notes that the texts elicited little immediate interest in Calcutta and London:

> At Paris, however, these Buddhist MSS. fell into the hands of Burnouf. Unappalled by their size and tediousness, he set to work, and it was not long before he discovered their extreme importance. After seven years of careful study, Burnouf published, in 1844, his "Introduction à l'Histoire du Buddhisme." It is this work which laid the foundation for a systematic study of the religion of Buddha. Though acknowledging the great value of the researches made in the Buddhist literatures of Thibet, Mongolia, China, and Ceylon, Burnouf showed that Buddhism, being of Indian origin, ought to be studied first of all in the original Sanskrit documents preserved in Nepal. Though he modestly called his work an "Introduction to the History of Indian Buddhism," there are few points of importance on which his industry has not brought together the most valuable evidence, and his genius shed a novel and brilliant light. The death of Burnouf in 1851 [*sic*] put an end to a work which, if finished according to the plan sketched out by the author in the preface, would have been the most perfect monument of oriental scholarship.[10]

Müller here summarizes a long and fascinating story about events that occurred over just eight years (1837–44) and that involved just two men, one English, one French; one, not in India, but Kathmandu, the other in Paris; two men who never met, two men who are rarely remembered today and almost never read, but who determined to a great extent the referent of the English word *Buddhism*. It is, at least to some degree, one of those forgotten tales from the annals of antiquarian scholarship that post-orientalist orientalists should regard with a certain degree of suspicion. However, it raises, at least implicitly, important questions for any history of the discourse of Buddhism and Science.

To set the stage, let me briefly describe the state of scholarly knowledge of Buddhism at the beginning of the nineteenth century. As described in chapter 2, the famous British Indologist Sir William Jones (1746–1794) had sought to establish a "Chronology of the Hindus" based on his study of Indian myths, particularly those about the incarnations of the god Viṣṇu. Jones, like other scholars of the day, assumed the accuracy of the "Mosaic Chronology," the history of the world set forth in the Bible, and thus was fascinated by Hindu myths of a great flood, which he saw as confirmation of the biblical account. He learned from Hindu pundits that the Buddha was (along with Kṛṣṇa and Rāma) an incarnation of Viṣṇu, and was revered for his condemnation of animal sacrifice. But Jones was unsure of where the Buddha had come from. In a lecture delivered in 1786, he speculated that the Buddha had come from Ethiopia, pointing to physiognomic similarities between Ethiopians and statues of the Buddha he had seen in India, especially the curly hair. Jones's theory of an African origin of the Buddha persisted into the first decades of the nineteenth century. In 1819 the French scholar Jean-Pierre Abel-Rémusat (1788–1832) published an article disputing Jones's view, entitled "Sur quelques épithètes descriptives de Bouddha qui font voir que Bouddha n'appartenait pas à la race nègre" (On Some Descriptive Epithets of Buddha Showing That Buddha Did Not Belong to the Black Race).[11]

The national origin of the Buddha, now identified as the founder of *Boudhism* (a term that first appears in English in 1801), was thus one of the topics that occupied the scholars of the day. Officers of the East India Company also surmised from statues and inscriptions that Buddhism had existed in India. Even at this early date, when so little was understood about Buddhism, some saw a profound difference between it and Hinduism. In a lecture describing the cave temples at Ellora, presented to the Literary Society of Bombay on November 2, 1813, William Erskine observed:

> The religion of the Bouddhists differs very greatly from that of the Brahmins; as in the latter, God is introduced everywhere,—in the former, he is introduced no where. The gods of the Brahmins pervade and animate all nature; the god of the Bouddhists, like the god of the Epicureans, remains in repose, quite unconcerned about human affairs, and therefore is not the object of worship. With them there is no intelligent divine being who judges of human actions as good or bad, and rewards or punishes them as such; —this indeed is practically the same as having no God. . . .
>
> As all the ideas of this religion relate to men, and as no incarnations or transformations of superior beings are recorded, it is obvious that in their temples we can expect to find no unnatural images, no figures

compounded of man and beast, no monsters with many hands or many heads.[12]

But these early scholars, amateur and professional, did not know when or how Buddhism arose. In addition to the question of where the Buddha had come from, one of the other great questions of the day thus was: which came first, Buddhism or Brahmanism (as Hinduism was then referred to)? The argument in favor of the priority of Buddhism was one based on reason: simplicity precedes complexity. Buddhism taught that the world was uncreated and that souls were mortal, whereas Brahmanism taught divine creation and the immortality of the soul. The primitive views of Buddhism could not have taken hold once the more advanced views of the brahmans were in place. Therefore, Buddhism must have come first.[13] A historical foundation was claimed for the opposite view. As one young scholar explained in 1836, "Buddhism (to hazard a character in a few words), is monastic asceticism in morals, philosophical scepticism in religion; and whilst ecclesiastical history all over the world affords abundant instances of such a state of things resulting from gross abuse of the religious sanction, that ample chronicle gives us no one instance of it as a primitive system of belief."[14]

That statement came from Brian Houghton Hodgson. He was born into a well-connected but impoverished Derbyshire family in 1801, one of seven children. At age fifteen he gained admission to Haileybury, the college that had been established by the East India Company in 1806 to train its future officials. He excelled at Bengali, Persian, Hindi, political economy, and classics. Following the standard curriculum of the company, after two years at Haileybury he went to the College of Fort William in Calcutta to continue his studies. Once in India, he immediately began to suffer liver problems and was told by a physician that his options were three: "six feet under, resign the service or get a hill appointment." He opted for the latter and was posted to a hill station in Kumaon, a region annexed by the British in the Himalayan foothills, before being assigned to Kathmandu as Assistant Resident to the Court of Nepal. As a consequence of the 1814–16 Gurkha War, the Nepalese court had been forced to accept a British Resident, but did so reluctantly, and restricted the movements of the British legation to the Kathmandu valley. With nothing better to do, Hodgson turned to Buddhism; although long dead in India, it still flourished in the Newar community of the Kathmandu valley. In a letter of August 11, 1827, to Dr. Nathaniel Wallich, Hodgson described how he came to undertake his studies:

> Soon after my arrival in Nipál (now six years ago), I began to devise means of procuring some accurate information relative to Buddhism: for, though the regular investigation of such a subject was foreign to my pursuits, my respect for science in general led me cheerfully to avail myself of the opportunity afforded, by my residence in a *Bauddha* country, for collecting and transmitting to Calcutta the materials for such investigation. There were, however, serious obstacles in my way, arising out of the jealousy of the people in regard to any profanation of their sacred things by an European, and yet more, resulting from Chinese notions of policy adopted by this government. I nevertheless persevered; and time, patience, and dexterous applications to the superior intelligence of the chief minister, at length rewarded my toils.
>
> My first object was to ascertain the existence or otherwise of *Bauddha* Scriptures in Nipál; and to this end I privately instituted inquiries in various directions, in the course of which the reputation for knowledge of an old *Bauddha* residing in the city of *Pátan*, drew one of my people to his abode. This old man assured me that Nipál contained many large works relating to Buddhism; and of some of these he gave me a list. Subsequently, when better acquainted, he volunteered to procure me copies of them. His list gradually enlarged as his confidence increased; and at length, chiefly through his kindness, and his influence with this brethren in the *Bauddha* faith, I was enabled to procure and transmit to Calcutta a large collection of important *Bauddha* scriptures.
>
> Meanwhile, as the *Pátan Bauddha* seemed very intelligent, and my curiosity was excited, I proposed to him (about four years ago) a set of questions, which I desired he would answer from his books. He did so; and these questions and answers form the text of the paper which I herewith forward. . . . Having in his answers quoted sundry slókas in proof of his statements; and many of the scriptures whence these were taken being now in my possession, I was tempted to try the truth of his quotations. Of that, my research gave me in general satisfactory proof. But the possession of the books led to questions respecting their relative age and authority; and, tried by this test, the *Bauddha's* quotations were not always so satisfactory. Thus one step led to another, until I conceived the idea of

> drawing up, with the aid of my old friend and his books, a sketch of the terminology and general disposition of the external parts of Buddhism, in the belief that such a sketch, though but imperfectly executed, would be of some assistance to such of my countrymen as, with the books only before them, might be disposed to enter into a full and accurate investigation of this almost unknown subject.[15]

Hodgson, age twenty-six at the time, employs the language of science. Finding himself in a Buddhist country, he determines to collect specimens of Buddhism and send them back to Europe for analysis. He describes here what would be his two most important contributions to Buddhist studies. The first is a paper he read in 1828, "Sketch of Buddhism, derived from the Bauddha Scriptures of Nipál." Although published at such an early age, it (together with a second paper from the same period[16]) was among the most widely read papers on Buddhism in the nineteenth century. It was noteworthy for several reasons, including that it was written, as Hodgson describes, in cooperation with "the old Bauddha," the old Buddhist, whose name was Amṛtānanda, the leading Newar scholar of Buddhism of the day. Hodgson's "Sketch" is all but forgotten now, and not entirely without reason, reasons to which we will return.[17]

Hodgson's other great deed was his collection and distribution of materials for the scientific investigation of Buddhism, in the form of Sanskrit manuscripts. As he reports, in 1824 he began accumulating Buddhist works in Sanskrit (and Tibetan[18]) and dispatching them around the world, beginning with the gift of 66 manuscripts to the library of the College of Fort William in 1827 and continuing until 1845: 94 to the Library of the Asiatic Society of Bengal, 79 to the Royal Asiatic Society, 36 to the India Office Library, 7 to the Bodleian Library of the University of Oxford, 88 to the Société Asiatique, and later 59 more to Paris. A total of 423 works were provided. They were largely ignored. The texts he sent to Calcutta were not even catalogued until 1882. Paris was the only destination where the Buddhist manuscripts received any attention.

Before turning to the reception and their recipient in Paris, we should briefly survey the state of Sanskrit studies and Buddhist studies there in the first decades of the nineteenth century. Charles Wilkins, a member of the East India Company and a founder of the Asiatic Society, had published *The Bhagavat-Geetâ; or, Dialogues of Kreeshna and Arjoon* in 1785. William Jones published *Sacontalá; or, The fatal ring: an Indian drama*, his translation of the Kālidasa's Sanskrit play *Śakuntala*, in 1792. Both were soon translated into French, but instruction in Sanskrit did not begin in Paris until 1803. In that year, the Peace of Amiens between Britain and France was broken, and Napoleon ordered the detention of all British males between the ages of eighteen and sixty then traveling in France. Among the 1,181 arrested was the Scottish lieutenant Alexander Hamilton (1762–1824), a former officer in the Bengal army who had come to Paris to examine an edition held by the Bibliothèque nationale of the *Hitopadeśa*, a famous Sanskrit collection of animal stories that Wilkins had also translated. Lieutenant Hamilton had spent the years 1783–95 in India working for the East India Company. He was an active member of the Asiatic Society, and he was a student of Sanskrit; upon his return from India to Scotland he became known as "Sanscrit Hamilton." Through the efforts of Count Constantine de Volney (1757–1820), Lieutenant Hamilton was paroled, and in return he offered to give private Sanskrit lessons to a small group in Paris. His students included the Count de Volney himself and the noted classicist Jean-Louis Burnouf (1775–1844), as well as a visitor from Germany, Friedrich Schlegel; Hamilton lived in Schlegel's house in Paris, and their studies of Sanskrit are said to have inspired Schlegel's *Über die Sprache und Weisheit der Indier* (1808).

In 1806 a student of Persian working as assistant librarian in the Department of Manuscripts at the Bibliothèque nationale, Antoine-Léonard de Chézy (1773–1832), determined to teach himself Sanskrit, using the disparate materials held by the library and the translations emanating from Calcutta. He succeeded brilliantly in his task, although his wife and children left him in the process. In 1814 the first chair of Sanskrit was established in Europe, not in Berlin or London but in Paris, at the Collège de France, and Chézy was named to hold it. Among his pupils was Eugène Burnouf, the son of one of Lieutenant Hamilton's students, and the man destined to receive Hodgson's gift.

Born in 1801, the same year as Hodgson, Burnouf learned Greek and Latin, and perhaps some Sanskrit, from his father. At the university, he excelled in both Sanskrit and Avestan, deciphering the manuscripts brought to France by Abraham Antequil-Duperron. He was a founder and an officer of the Société Asiatique, established in 1822. In 1826 he published his first work to touch on Buddhism, *Essai sur le pâli*.

At age twenty-eight, he was named professor of general and comparative grammar at the École normale. While there, he received an award from Count de Volney for his work in "the transcription of Asiatic scriptures in Latin letters." When his teacher Chézy died in the cholera epidemic of 1832, Burnouf was named to replace him in the Sanskrit chair at the Collège de France. Hodgson's manuscripts arrived five years later, and Burnouf immediately turned his attention from the study of Hindu texts to the study of Buddhist texts, which occupied him until his untimely death in 1852.

During the first decades of the nineteenth century, Buddhist scholarship occupied a small domain in Europe. But it was a domain populated by the pioneers of the field of oriental philology. These included the great French sinologist Jean-Pierre Abel-Rémusat (1788–1832), who translated the account of the Chinese monk Faxian's fifth-century pilgrimage to India; Stanislas Julien (1797–1873), who translated the life of Xuanzang; the German Mongolist Julius von Klaproth (1783–1835), who published a life of the Buddha in 1823; the Norwegian scholar Christian Lassen (1800–1876), with whom Burnouf collaborated in their *Essai sur le pâli*; the Dutch scholar Isaak Jakob Schmidt (1779–1847), who published a German translation of the *Diamond Sūtra* from the Tibetan in 1837; the Transylvanian Alexander Csoma de Körös (1784–1842), who studied Tibetan texts in Ladakh; and George Turnour (1799–1843), a British civil servant in Sri Lanka, who published in the *Ceylon Almanac* in 1833 and 1834 a work entitled *Epitome of the History of Ceylon, and the Historical Inscriptions*. This contained a translation of "the first twenty chapters of the Mahawanso and a prefatory essay on Pali Buddhistical literature." Significantly, all of these scholars were working on texts from countries other than India, and in languages other than Sanskrit: Chinese, Mongolian, Tibetan, Burmese, Pāli. Thus, Eduard Roer (1805–1866), surveying European knowledge of Buddhism in an 1845 review, noted that the initial understanding of Buddhism in Europe had come from "secondary sources," leading him to observe, "Our first acquaintance with Buddhism was in fact not a kind to invite research; the mixture of extravagant fables, apparent historical facts, philosophical and religious doctrines was so monstrous, that it seemed to defy every attempt to unravel it."[19]

As noted earlier, Brian Hodgson had sent a parcel of Sanskrit manuscripts from Kathmandu to the Société Asiatique in Paris; they arrived in 1837, and Burnouf immediately began reading, a skill in which he was perhaps unmatched in his day. He notes in passing, "I can assert that there is nothing in all Sanskrit literature as easy to understand as the texts of Nepal, apart from some terms the Buddhists used in a very special way; I shall not give any proof other than the considerable number of texts it has been possible for me to read in a rather limited time."[20] The "considerable number of texts," it turns out, were the eighty-eight manuscripts from Hodgson, a corpus that included many lengthy and (at least in the estimation of lesser mortals) difficult sūtras. One must also acknowledge Burnouf's remarkable dedication, which placed him at his desk at 3:00 AM each morning—a practice that his contemporaries blamed for his early death.

This initial group of eighty-eight (Hodgson later sent Burnouf an additional fifty-nine titles) included sūtras and tantras of Sanskrit Buddhism, composed for the most part during the first six centuries of the common era, largely lost in India but preserved in Nepal; works that in India, and in translations into Chinese and Tibetan, were among the most important in the history of Buddhism. To list just ten, Burnouf received the *Aṣṭasahāsrikāprajñāpāramitā* (the Perfection of Wisdom in Eight Thousand Lines), one of the earliest and most influential of the perfection of wisdom (*prajñāpāramitā*) texts; the *Gaṇḍavyūha*, regarded as the Buddha's most profound teaching by the Huayan schools of East Asia; the *Sukhāvatīvyūha*, the fundamental sūtra for the Pure Land traditions; the *Laṅkāvatāra*, a central text for the Yogācāra school in India and the Chan and Zen traditions of East Asia; the *Lalitavistara*, a baroque account of the Buddha's early life; the *Guhyasamāja*, among the most influential of Buddhist tantras; the *Abhidharmakośa*, Vasubandhu's important compendium of doctrine; the *Bodhicaryāvatāra*, an eighth-century poem by Śāntideva on the practice of the bodhisattva; the *Buddhacarita*, Aśvaghoṣa's second-century life of the Buddha; and the *Saddharmapuṇḍarīka*, the famous *Lotus Sūtra*.

Burnouf made a preliminary survey of these texts and identified one as the most interesting: the *Lotus Sūtra*, a choice that strikes us today as uncanny, for Burnouf did not know that it is perhaps the single most influential text in the history of Buddhism. He wrote to Hodgson on June 5, 1837, that he was spending all of his spare time reading it. He completed a translation of the entire sūtra into French two years later and had the book typeset. The translation ran to 283 printed pages, with 150 pages of notes and 433 pages of appendices. However, before it was published, he felt he must first compose a brief introduction to the fundamental doctrines of Buddhism for the scholars of Europe. Although he retained the word *introduction* in the title of his composition, it was not brief. *Introduction à l'histoire du Buddhisme indien*, published in 1844, is 653 large pages long. It was intended as the first of as many as four volumes that would precede the publication of *Lotus Sūtra*. The second volume, in five memoranda (as Burnouf calls them), would survey the literature of Pāli Buddhism, as the first volume does the literature of Sanskrit Buddhism. The third volume would then compare the Sanskrit collection of Nepal and the Pāli collection of Sri Lanka. Burnouf was convinced "that the fundamental and truly antique elements of Buddhism must be sought in what the two Indian redactions of the religious books, that of the North which uses Sanskrit and that of the South which uses Pāli, will have kept in common."[21] It was also his expectation that the difference between the two collections would be less in the content than in the form and the classification of the books that formed those collections. The fourth volume, which he sometimes describes as a "historical sketch," would trace the chronology of Indian Buddhism, seeking first to establish the date of the Buddha's death and then the chronology of the various councils that Burnouf regarded as so critical for understanding the formation of the Buddhist canons. But Burnouf died of kidney failure in 1852, six weeks after his fifty-first birthday. Among his papers were found an almost complete translation into French of the *Aṣṭasahāsrikāprajñāpāramitā* (the Perfection of Wisdom in Eight Thousand Lines), a translation of the *Kāraṇḍavyūha*, an important *sūtra* about Avalokiteśvara, and over a thousand pages of translations from the *jātaka*s, the stories of the Buddha's former lives. His translation of the *Lotus Sūtra* was published posthumously.

The work that Burnouf considered only an introduction to a multivolume project, only an orientation prior to a long textual excursion, was destined, or doomed, to be his most influential work. The impact of the *Introduction* was immediate, and it extended well beyond France, and beyond the infant discipline of Buddhist studies. In America it was read by Ralph Waldo Emerson and Henry David Thoreau.[22] On May 28, 1844, the year of the *Introduction*'s publication, Yale Sanskrit instructor Edward Eldridge Salisbury (1814–1901), a Congregationalist deacon and student of Burnouf's, delivered a lecture entitled "Memoir on the History of Buddhism" at the inaugural meeting of the American

Oriental Society. This fifty-page report, based largely on Burnouf's work (and eventually published in the *Journal of the American Oriental Society* in 1849) is the first scholarly article on Buddhism written in North America.

Burnouf's *Introduction* was read in Germany by Schelling (who praised it for refining his understanding of nirvāṇa and noted how remarkable it was that France, with its political instability, could produce a man like Burnouf), by Schopenhauer,[23] and by Nietzsche. Wagner wrote, "Burnouf's *Introduction to the History of Indian Buddhism* interested me most among my books, and I found material in it for a dramatic poem, which has stayed in my mind ever since, though only vaguely sketched."[24] As noted in chapter 2, the material for this poem came specifically from Burnouf's description of the *Śārdūlakarṇāvadāna*. Wagner's Buddhist opera, *Die Sieger*, although listed in the timetable he presented to King Ludwig II, was unfortunately never completed.

It can be argued that Burnouf's *Introduction à l'histoire du Buddhisme indien* is the single most important work in the history of the academic study of Buddhism. After its initial publication in 1844, however, it was reprinted only once, in 1876. One could argue that the book has all but disappeared and today remains unread and unexamined—not because it is outdated or has been superseded, but because it became so fully integrated into the mainstream representation of Buddhism, which it created, that it is no longer visible.

What Burnouf published in 1844 would provide the foundation for the study of Indian Buddhism for the next century. He provides lengthy discussions of the discourses of the Buddha, the sūtras; of the code of monastic conduct, the vinaya; and of the metaphysical treatises, the Abhidharma. Burnouf offers extended passages translated from a great variety of texts, including many *avadāna*s, tales of the former lives of the Buddha and his disciples. There are essays on topics that continue to draw the attention of scholars, such as the meaning of terms like *nirvāṇa* and *pratītyasamutpāda* (dependent origination). There are also discussions of Sanskrit terms for weights and measures and on varieties of sandalwood.

Near the end of the volume he takes up the issue of Buddhist tantra. The tone of his estimation of the tantras would persist for more than a century. He wrote, "It is not my intention to long dwell on this part of the Nepalese collection, which I am inclined to regard as the most modern of all, and whose importance for the history of human superstitions does not compensate for the mediocrity and the vapidity. It is certainly not without interest to see Buddhism, which in its first organization had so little of what makes a religion, end in the most puerile practices and the most exaggerated superstitions."[25] That is, the original teachings of the Buddha, which may not be properly labeled a "religion," were eventually polluted with the most primitive elements of Hinduism. This also is a view that would long persist.

So great was the influence of Burnouf's tome, that judgments of its accuracy have been based on standards of accuracy that Burnouf himself set and a portrait of Buddhism that Burnouf himself painted. Perhaps his most beautifully drawn, and enduring, of the portraits that appear in his magnum opus is that of the Buddha.

> [There] was born, in a family of *kṣatriyas*—that of the Śākyas of Kapilavastu, who claimed descent from the ancient solar race of India—a young prince who, renouncing the world at the age of twenty-nine, became a monk under the name of *Śākyamuni* or also *śramaṇa Gautama*. His doctrine, which according to the sūtras was more moral than metaphysical, at least in its principle, rested upon an opinion accepted as a fact and upon a hope presented as a certitude. This opinion is that the visible world is in perpetual change; that death succeeds life and life death; that man, like all that surrounds him, revolves in the eternal circle of transmigration; that he successively passes through all forms of life from the most elementary to the most perfect; that the place he occupies on the vast scale of living beings depends upon the merit of the actions he performs in this world; and thus the virtuous person must, after this life, be reborn with a divine body, and the guilty with a body of the damned; that the rewards of heaven and the punishments of hell have only a limited duration, like everything in the world; that time exhausts the merit of virtuous actions as it effaces the faults of evil actions; and that the fatal law of change brings the god as well as the damned back to earth, in order to again put both to the test and make them pass through a new series of transformations. The hope that Śākyamuni brought to humanity was the possibility to escape from the law of transmigration, entering what he calls nirvāṇa, that is to say, annihilation. The definitive sign of this annihilation was death; but a precursory sign in this life announced the men predestined for this supreme liberation; it was the possession of an unlimited science,

> which gave him a clear view of the world, as it is, that is to say, the knowledge of physical and moral laws; and in a word, it was the practice of the six transcendent perfections: that of alms-giving, ethics, science, energy, patience, and charity. The authority upon which the monk of the Śākya race supported his teaching was entirely personal; it was formed of two elements, one real and the other ideal. The first was the consistency and the saintliness of his conduct, of which chastity, patience, and charity formed the principal features. The second was the claim he made to be buddha, that is to say, enlightened, and as such to possess superhuman science and power. With his power, he performed miracles; with his science, he perceived, in a form both clear and complete, the past and the future. Thereby, he could recount everything that each person had done in his previous existences; and so he asserted that an infinite number of beings had long ago attained like him, through the practice of the same virtues, the dignity of buddha, before entering into complete annihilation. In the end, he offered himself to humanity as its savior, and he promised that his death would not annihilate his doctrine; but that this doctrine would endure for a great number of centuries after him, and that when his salutary action ceased, there would come into the world a new buddha, whom he announced by name and whom, before descending to earth, the legends say, he himself had crowned in heaven, with the title of future buddha.[26]

Unusual in its day for the eloquence and confidence of its expression, Burnouf's description seems today conventional in its content. This is the Buddha that we know. Burnouf even endows the Buddha with science, his translation of the Sanskrit term *prajñā* (usually rendered in English today as "wisdom"). It is important to recall that the Buddha had never previously been described in quite these terms by a European, or by an Asian, for that matter. Burnouf does not deviate from this view of the Buddha, across the 653 pages of his text.

For Burnouf, the importance of the Buddha is to be found in his preaching of what the French scholar calls "pure morality"; the extravagant metaphysics of the *prajñāpāramita*, and certainly the tantras, are inventions of a later age. Throughout the sūtras, the Buddha is above all human, and the power of his humanity was such that it could overthrow the great weight of culture. Burnouf writes, "He lived, he taught, and he died as a philosopher; and his humanity remained a fact so incontestably recognized by all that the compilers of legends whom miracles cost so little did not even have the thought of making him a god after his death."[27] Or, as he puts it elsewhere, "This respect for human truth in Buddhism, which prevented the disciples of Śākya from transforming the man into God, is quite remarkable for a people like the Indians, among whom mythology has so easily taken the place of history."[28]

Burnouf did not turn a blind eye to the miraculous elements he encountered in his reading. He devotes thirty-three pages, for example, to the story of the miracle at Śrāvastī, where the Buddha famously rose into the air and simultaneously emitted fire and water from his body. He is careful to note, however, that the Buddha performs miracles only when he is challenged by the brahmans. And he takes a certain delight in their humiliation and defeat. The brahmans, who appear often in Burnouf's book, seem to function for him as Indian Jesuits. But in the end he is not interested in miracles. Describing the Buddha's means of converting people to the dharma, he writes, "These means were the teaching and, according to the legends, the miracles. Let us leave the miracles aside for the moment, which are no more worthy than those with which the brahmans opposed him. But the teaching is a means quite worthy of attention and which, if I am not mistaken, was unheard of in India before the coming of Śākya."[29]

Burnouf's preference for the style and the language of what he called "the simple sūtras" over that of what he called "the developed sūtras" (that is, the Mahāyāna sūtras) was not simply a matter of taste, although it was also that. He discerned in the difference a historical key that would open the door to what he regarded as perhaps the most important question in his endeavor: among the thousands of pages of manuscripts he had received from Hodgson, which represented the Buddha's original teaching, unadulterated by the tradition? Thus, he writes:

> The ordinary sūtras show us Śākyamuni Buddha preaching his doctrine in the midst of a society that, judging from the legends in which he plays a role, was profoundly corrupt. His teaching is above all moral; and although metaphysics is not forgotten, it certainly occupies a less grand position than the theory of virtues imposed by the law of the Buddha, virtues among which charity, patience, and chastity are without objection at the first rank. The law, as Śākya calls it, is not set forth dogmatically in these books; it is only mentioned there, most often in a vague manner, and presented in its applications rather than in its principles. In order to

> deduce from such works a systematic exposition of the belief of the Buddhists, it would be necessary to have a very great number of them; still, it is not certain that one would be able to succeed in drawing a complete picture of Buddhist morality and philosophy by this means; for the beliefs appear there, so to speak, in action, and certain points of doctrine recur there on each page, while others are hardly mentioned, or not at all. But this circumstance, which for us is a true imperfection, also has its advantages from the historical perspective. It is a certain index of the authenticity of these books, and it proves that no systematic effort attempted to complete them afterwards, nor to place them, through later additions, at the level of progress that Buddhism certainly reached in the course of time. The developed sūtras have, as far as doctrine is concerned, a marked advantage over the simple sūtras, for the theory there proves to be more advanced from the dual perspective of dogma and metaphysics; but it is precisely this particularity which makes me believe that the *vaipulya* sūtras are later than the simple sūtras. These latter make us witness to the birth and first developments of Buddhism; and if they are not contemporary with Śākya himself, they at least have preserved for us the tradition of his teaching very faithfully.[30]

Burnouf did not live to write his second volume on the Pāli collection of Ceylon. But if he had, there is little doubt that the Pāli *sutta*s would fit easily into the "simple sūtra" category, and would be judged as those closest to the Buddha, as the German scholar Hermann Oldenberg would do in the decades after Burnouf's death. Burnouf's image of the Buddha would remain that of a man of the Enlightenment, in every sense of the word.

At the height of Europe's rage for Sanskrit, Burnouf found the Sanskrit Buddha, and defined him for the century to come. From that point on, Sanskrit would be the medium through which Buddhism must be understood, and the true Buddha would be the Buddha of the texts, texts from a land where Buddhism had been dead for centuries. And this Buddha would be the Buddha of Burnouf, a mortal and moral philosopher who offered his teachings, with their message of freedom from suffering, to all members of society. This Buddha would become the Buddha of Buddhism and Science.

Burnouf's legacy is safe, even if his book is unread. But what of Hodgson, to whom Burnouf dedicated his translation of *Le Lotus*? Burnouf showed little love for what he called metaphysics, although he dutifully devoted a large section of the *Introduction* to the Abhidharma. Philosophy, however, is what interested Hodgson most about Buddhism, and in his 1828 "Sketch of Buddhism" he offered what he regarded as his most important contribution, "the distinction of the various schools of philosophy; the peculiar tenets of each school." These were the four major schools of Buddhist philosophy that had been described to him by the Newar pundit Amṛtānanda: the Swábhávika (with its subschools, the simple Swábhávika and the Prájnika Swábhávika), the Aishwarika, the Kármika, and the Yátnika. For the non-orientalist interested in Buddhist philosophy in the nineteenth century, Hodgson's delineation of the four major schools of Buddhist thought proved authoritative; one finds these terms repeated in any number of expositions of Buddhism, including some influential textbooks, as well as in theological treatises of another variety, such as Madame Blavatsky's *The Secret Doctrine*. The four schools and their doctrines thus generated considerable fascination. Yet what is perhaps more fascinating for our purposes is that there is no evidence that schools with these names ever existed in India, nor were the doctrines ascribed to them coherent. When they were unable to find references to these schools in the treatises of Indian Buddhism (which had been Hodgson's claim), European scholars of Buddhism assumed that the four systems were the schools of Nepalese Buddhism. This was also wrong, but such was Hodgson's authority that references to the four schools of Nepalese Buddhism persisted into the twentieth century.

Brian Houghton Hodgson left two legacies: his legacy as a collector and his legacy as an interpreter, his legacy as someone who got things right and his legacy as someone who got things wrong. Neither of these legacies would have been possible, it must be emphasized, without the work of Amṛtānanda, Hodgson's "native informant" *avant la lettre*. Their collaboration raises a number of difficult questions about the European scholar's relation to the Buddhist text and to the living Buddhist. Amṛtānanda, unlike so many of the pundits who were so invaluable to other officials of the East India Company, was not simply an anonymous informant: Hodgson (at least by his own account) engaged in a dialogue with Amṛtānanda, repeatedly praised his learning, and mentioned him by name. But Amṛtānanda misinformed Hodgson, or Hodgson misunderstood Amṛtānanda.

From his post in Kathmandu, Hodgson remained abreast of the growing scholarship on Buddhism emanating from the universities of Europe, scholarship that he sometimes regarded with an unconcealed contempt, referring to its authors as "closet students." At the same time, amid the defense of his own method, and his testimony to the learning of Amṛtānanda, he could not entirely cede authority on Buddhism to Buddhists. Hodgson then found himself in a difficult position, suspected by those whose respect he sought, suspicious of those upon whom his authority rested—a position that led him essentially to abandon his study of Buddhism before he was thirty-five, turning his attentions to linguistics and ornithology. He was thus led to defend himself with a certain ambivalence: "Let the closet student, then, give reasonable faith to the traveller, even upon this subject; and, whatever may be the general intellectual inferiority of the orientals of our day, or the plastic facility of change peculiar to every form of polytheism, let him not suppose that the living followers of Buddha *cannot* be profitably interrogated touching the creed they live and die in."[31]

We are left, then, with an apparent conclusion that confounds our expectations, and perhaps our hopes. Brian Hodgson, living in the heart of a Buddhist community, engaged in dialogue, in the vernacular, with leading Buddhist scholars, got it wrong. Eugène Burnouf, never leaving Europe, rarely leaving Paris, never seeing a Buddhist, much less conversing with one, sitting instead alone in his study surrounded by Sanskrit manuscripts copied for him by Hodgson's Newars, got it right. Assuming for a moment that this impression is correct, how can we explain it? It was Hodgson's fate to live a long life, a life that saw sweeping changes in the production of knowledge. Over the course of the nineteenth century, authority would shift away from the amateur and to the professional scholar, away from the expert collector and to the academic local archivist, away from the traveler and to the professor, away from the enthusiast and to the scientist. If the European study of Buddhism is an academic discipline (which remains a question), and if that discipline has a founder, it is Eugène Burnouf and not Brian Hodgson. But Burnouf was only able to make such remarkable strides in his understanding of Buddhism because he had before him in Paris the texts that Hodgson had dispatched from Kathmandu. It was Hodgson's dispatch (which, in turn, would not have been possible without Amṛtānanda) that helped to make the science of Buddhism possible.

Burnouf, the son of a classicist, seems to have been trained almost from birth to decipher dead languages. When he received eighty-eight manuscripts in a dead language, manuscripts of works composed in a land where Buddhism was long absent, Burnouf could set to work shielded from the sensations of a Buddhist setting. Left only with his prodigious Sanskrit skills, his dogged analysis, and his imagination of what must have been, he created a historical narrative of Buddhism—from pristine origin, to baroque elaboration, to degenerate decline—based entirely on his reading of a random group of texts that arrived on his desk, as if from nowhere. Burnouf's narrative would define the Buddha and Buddhism for the world, and would outline an agenda of knowledge whose effects we continue to feel today, effects that, in the language of Buddhism, are either pleasurable, painful, or neutral. How one would classify those effects depends, of course, on one's perspective, a perspective that in large part remains that of Burnouf.

Burnouf died young. The father of the science of Buddhism did not live to see the rise of the Buddhism and Science discourse. His legacy, however, was carried on by his students, including Max Müller, who did. During the last half of the nineteenth century, interest in Buddhism had spread to the European and American publics, spurred by such works as Edwin Arnold's poem about the life of the Buddha, *The Light of Asia*. The Theosophical Society, which claimed the Buddha among its mahatmas, was thriving in Europe and America. Meanwhile, the scholarly or "scientific" study of Buddhist texts progressed steadily. There were many, then, who laid claim to the Buddha and his legacy, claims that sometimes led to contestation. In 1893, five years after he and Colonel Olcott had agreed to disagree, Müller again did verbal battle, this time in print, with the Theosophists.

Over the course of four months, from May to August 1893, an exchange took place in the pages of the *Nineteenth Century*. It was an exchange between Max Müller, the orientalist who, like so many members of the guild, had never been to the Orient, and A. P. Sinnett, the Theosophist, proudly not an orientalist, who had returned from eleven years in India.

Müller published "Indian Fables and Esoteric Buddhism," a lengthy essay, in the May 1893 issue of the periodical. By this time, at age sixty-nine, he was at the height of his fame. He was president of the International Congress of Orientalists and had politely declined an invitation to preside at the World's Parliament of Religions in Chicago. He was receiving letters and telegrams from around the world, congratulating him on the jubilee of his Doctorate of Philosophy. He had given the Gifford Lectures in Natural Theology an unprecedented four times, and in 1893 was preparing the fourth set of lectures, "Theosophy or Psychological Religion," for publication. About the choice of title, Müller wrote: "The venerable name, so well known among early Christian thinkers, as expressing the highest conception of God within the reach of the human mind, has of late been so greatly misappropriated that it was high time to restore it to its proper function. It should be known once and for all that one may call oneself a theosophist without . . . believing in any occult sciences or black art."[32]

In the May issue of the *Nineteenth Century*, Müller begins by recounting at some length the various traveler's tales, fables, and outright hoaxes that had been duly recorded as fact by credulous Europeans, from Plato to Sir William Jones. This leads him eventually to the subject of his essay.

> A very remarkable person, whose name has lately become familiar in England also, felt strongly attracted to the study of Buddhism. I mean, of course, the late Madame Blavatsky, the founder of Esoteric Buddhism. I have never met her, though she often promised, or rather threatened, she would meet me face to face at Oxford. She came to Oxford and preached, I am told, for six hours before a number of young men, but she did not inform me of her presence. At first she treated me almost like a Mahâtma, but when there was no response I became, like all Sanskrit scholars, a very untrustworthy authority.[33]

He goes on to describe Madame Blavatsky as "a clever, wild, and excitable girl"[34] who became interested in Indian philosophy "through the dark mists of imperfect translations" before turning to Buddhism. But, he explains, "No one can study Buddhism unless he learns Sanskrit and Pâli, so as to be able to read the canonical books, and at all events to spell the names correctly. Madame Blavatsky would do neither, though she was quite clever enough, if she had chosen, to have learnt Sanskrit or Pâli."[35] He notes her deficiencies in the relevant languages repeatedly, suggesting that her informants in India were equally deficient. She shrewdly exempted herself from all scholarly critique by declaring that hers was not the Buddhism that was known to the world, but instead an Esoteric Buddhism, one that preceded both Brahmanism and Christianity. This Buddhism she claimed to have learned from enlightened beings in Tibet, "quite safe," Müller notes, "from any detectives or cross-examining lawyers."[36] But in fact, he concludes, "There is nothing that cannot be traced back to generally accessible Brahmanic and Buddhistic sources, only everything is muddled or misunderstood. If I were asked what Madame Blavatsky's Esoteric Buddhism really is, I should say it was Buddhism misunderstood, distorted, caricatured. There is nothing in it beyond what was known already, chiefly from books that are now antiquated. The most ordinary terms are misspelt and misinterpreted."[37]

He continues at some length to describe the historical relation of Buddhism and Brahmanism, one of the venerable themes of oriental scholarship in the first half of the nineteenth century, concluding with the observation that one of the great differences between the two was to be found on the question of secrecy. He concludes that "whatever was esoteric or secret was *ipso facto* not Buddha's teaching; whatever was Buddha's teaching was *ipso facto* not esoteric."[38] And he quotes passages from the Pāli canon in support of this claim. He thus finds it highly ironic that it was Buddhism, among all the other religions, that Madame Blavatsky selected as being somehow "esoteric."

Müller does not launch this attack against Madame Blavatsky merely in the defense of the principles of scholarship, however.

> It is because I love Buddha and admire Buddhist morality that I cannot remain silent when I see his noble figure lowered to the level of religious charlatans, or his teaching misrepresented as esoteric twaddle. I do not mean to say that Buddhism has never been corrupted and vulgarised when it became the religion of barbarous and semi-barbarous people in Tibet, China, and Mongolia; nor should I wish to deny that it has in some places been represented by knaves and impostors as something mysterious, esoteric, and unintelligible.[39]

Indeed, Müller concedes, there are Sanskrit Buddhist texts whose titles contain the term *guhya*, "secret," in Sanskrit, and there are many Buddhist texts that remain entirely unknown, such that, in fact, "we know as yet very little, and that we see but darkly through the immense mass of its literature and the intricacies of its metaphysical speculations."[40] Madame Blavatsky's crime, then, is not that she claims that there is much about Buddhism that is not widely known, for this is indeed the case, but that she does not provide us with the titles of the texts from which she derives her knowledge. It is, however, unnecessary to go to Madame Blavatsky or her mahatmas to dispel our ignorance about Buddhism. Instead, Müller advises, "We should go to the manuscripts in our libraries, even in the Bodleian, in order to do what all honest Mahâtmas have to do, copy the manuscripts, collate them, and translate them."[41]

Müller's essay provides a fascinating perspective on the Buddhism of the orientalists at the end of the nineteenth century and, indeed, on the Buddhism of Buddhism and Science. But let us refrain from such analysis for the time being to introduce Müller's interlocutor.

As mentioned above, Blavatsky and Olcott had sailed to India in 1879 and arrived in Bombay, where they proclaimed themselves to be Hindus. They proceeded the next year to Ceylon, where they both took the vows of lay Buddhists. Upon their return to India, they traveled north, where they met Alfred Percy Sinnett (1840–1921). Sinnett was a journalist, having worked at the *Hong Kong Daily Press* and the *Evening Standard* before going to India in 1872 to become editor of *The Pioneer* in Allahabad, one of the major newspapers in India. (One of his reporters was Rudyard Kipling.) During a visit to London in 1875, Sinnett attended a séance at the house of the renowned medium Mrs. Guppy (Agnes Elizabeth Guppy, first made famous by Alfred Russel Wallace) and became fascinated with spiritualism. He later read Madame Blavatsky's first major work, *Isis Unveiled*, and his newspaper covered Blavatsky and Olcott's travels in India. In 1880 Blavatsky and Olcott accepted an invitation to visit the Sinnett home in the hill station of Simla in northern India, where they remained for six weeks. After a demonstration of paranormal powers by Madame Blavatsky, Sinnett asked her to place him in contact with a mahatma. The first mahatma she approached initially refused, but the second agreed, and between 1880 and 1884, Sinnett carried on a prodigious correspondence with the two most famous mahatmas, the Master Koot Hoomi (KH) and the Master Morya (M). His letters formed the basis of three important works in the Theosophical canon: *The Occult World* (1881), *Esoteric Buddhism* (1883), and *The Mahatma Letters to A. P. Sinnett from the Mahatmas M. & K. H.* (1923).

By May 1893, Madame Blavatsky was dead and A. P. Sinnett had been back in London for ten years; he had been dismissed from his position at *The Pioneer* shortly after the publication of his first book on Theosophy. Upon his return to England, he was disappointed to have been passed over for the presidency of the London Lodge of the Theosophical Society; a letter from Koot Hoomi himself had encouraged the membership to support his rival, Anna Kingsford.

In June 1893, the editor of the *Nineteenth Century* published Sinnett's reply to Müller's essay. After noting the absurdity of Müller's claim "that Buddhism cannot contain any teaching hitherto kept secret, because the books hitherto published do not disclose any secrets of the kind,"[42] Sinnett turns not to a defense of Madame Blavatsky but to a defense of himself. Müller is quite wrong to ascribe the formulation of the system of Esoteric Buddhism to Madame Blavatsky. Thus, before he can "vindicate the ideas he [Müller] seeks to disparage," Sinnett must first set the record straight.

> In 1883 I was enabled to bring into intelligible shape a view of the origin and destinies of man derived from certain teachings with which I was favoured while in India. It challenged the attention of Western readers because it seemed to furnish a more reasonable interpretation of man's spiritual constitution and of the world's purpose, than any with which European thought had previously been concerned. It provided something like a scientific abstract of all religious doctrine, by the help of which it was easy to separate the wheat from the chaff in various ecclesiastical creeds. Allowing for symbolical methods of treatment as entering largely into popular religions, the new teaching showed that Brahmanism, Buddhism, and Christianity could be accounted for as growing up at various periods in India and Europe from the same common root of spiritual knowledge. But since Buddhism had apparently separated itself less widely than other religions from the parent stem, I gave my book the title *Esoteric Buddhism*, partly in loyalty to the exterior faith preferred by those from whom my information had come, partly because even in its exterior form that religion was already attracting a great deal of sympathetic interest in Europe, and seemed the natural bridge along which European thinking might be conducted to an appreciation of the beautifully coherent and logical view of Nature I had been enabled to obtain.[43]

Sinnett puts forth the basic Theosophical tenet that the great religious traditions of the world all sprang forth from the same root of esoteric wisdom. Among the exoteric versions of the world's religions, Buddhism remains closest to this ancient source. He also concedes that the popularity of Buddhism in the West might lead people eventually to Theosophy. Hence, he decided to entitle his book *Esoteric Buddhism*.

Sinnett goes on to explain that Madame Blavatsky's aim, especially in *Isis Unveiled*, was not "to teach anything in particular, but to stir up interest in an unfamiliar body of occult mysteries."[44] It was Sinnett himself, however, who "was entrusted with the task of putting into intelligible shape the views of life and nature entertained by certain Eastern initiates who were interested in the Theosophical Society."[45] Müller, it seems, is unaware of all this.

Müller is further mistaken in claiming that nothing of the secret teachings is present in the sacred books of the Buddhists. It is there, but visible only to those who have received instruction in esoteric doctrine. And to prove his point, Sinnett explains the esoteric meaning of the Buddha's last meal.

Before proceeding to Sinnett's exegesis, it should be noted that precisely what it was that the Buddha ate before he passed into nirvāṇa is a question that has puzzled both monastic commentators and academic exegetes for two millennia. The dish is specified in the Pāli canon with the compound *sūkaramaddava*, composed of the word for "pig" and the word for "soft." It is unclear whether this means something soft that is consumed by pigs, such as a type of mushroom or truffle, or perhaps bamboo shoots that had been trampled by pigs. On the other hand, the compound could be the name of some kind of pork dish. The Indian and Sinhalese commentators prefer, although not unanimously, the latter interpretation.

Sinnett refers to this view, given currency by Thomas W. Rhys Davids, as a "ludicrous misconception." (He had earlier expressed his contempt for the views of Rhys Davids—the leading British scholar of Pāli Buddhism—on the doctrine of rebirth in the final chapter of *Esoteric Buddhism*.) He explains that "Common-sense ought to have been startled at the notion that the diet of so ultra-confirmed a vegetarian as a Hindoo religious teacher could not but be, could be invaded by so gross an article of food as roast pork. But worshippers of the letter which killeth are apt to lose sight of common-sense."[46] One might assume from this that Sinnett allies himself with the truffles camp. However, he offers another explanation.

He asks the reader to recall that in the *Viṣṇu Purāṇa*, the god Viṣṇu took the form of a boar in one of his incarnations and lifted up the world with his tusks in order to rescue it from a great flood. In the account of the Buddha's last days, boar's flesh thus symbolizes esoteric knowledge that has been prepared for popular consumption by the multitudes. The Buddha had attempted to bring such knowledge to the populace and had died as a result. Sinnett finds support for his reading in the details of the story. The Buddha indeed instructs his host Cunda to serve this dish to him alone, not to his monks, and to bury the rest. Although there are traditional explanations for this request, Sinnett takes it to mean that "no one of lesser authority than himself must take the responsibility of giving out occult secrets."[47]

The remainder of Sinnett's response is devoted to chastising Müller for judging the message (Theosophy) by the apparent messenger (Madame Blavatsky and the mahatmas). He then presents a summary of Theosophical doctrine; because Müller did not see fit to provide one, it is left to Sinnett to do so. Theosophy, he writes, "gives us religion in the form of abstract spiritual science which can be applied to any faith, so that we may sift its crudities from its truth."[48] And he concludes with a prophecy:

> Every advance of knowledge leaves some people aground in the rear, and there are hundreds of otherwise distinguished men amongst us who will probably never in this life realise the importance of new researches on which many other inquirers besides theosophists are now bent. But their immobility will be forgotten in time. Knowledge will advance in spite of them, and views of nature, at first laughed at and discredited, will be taken after a while as matters of course, and, emerging from the shadow of occultism, will pass down the main current of science. Those of us who are early in the field with our experience and information would sometimes like to be more civilly treated by the recognized authorities of the world; but that is a very subordinate matter after all, and we have our rewards, of which they know nothing. We are well content to be in advance even at the cost of some disparaging glances from our less fortunate companions.[49]

Like Olcott before him, but somewhat more caustically, Sinnett portrays Müller as someone satisfied with a superficial knowledge, and who is thus remaining blind to the deeper and authentic meaning of the texts he reads. Sinnett portrays himself as something of a prophet, not of a new religion but of a new science. He prophesies a transformation in human knowledge, one in which what once was dismissed as "the occult" will enter the mainstream in a new paradigm as science and spirituality converge. Sinnett is content to suffer the mockery of Müller, knowing that soon enough the likes of Müller will be left behind.

Müller responded in the August issue of the *Nineteenth Century*, expressing his surprise that Sinnett would claim credit for Esoteric Buddhism so soon after Madame Blavatsky's death. He apologizes for not having read Sinnett's books, but explains that his original essay was about Blavatsky, not Sinnett; he had only written about her in the first place because of all the appeals he had received to do so. He finds Sinnett's statement (which Müller paraphrases as), "Whether I obtained Esoteric Buddhism from a Mahâtma on the other side of the Himalaya or from my own head is of no consequence,"[50] to be ominous. Sinnett must provide some evidence that such teachings can be found somewhere in Tibet. But if he wants Theosophy to be judged simply on the basis of its doctrine, Müller confesses that he finds Sinnett's summary to be incomprehensible.

On the Buddha's last meal, Müller reveals to Sinnett that Sanskritists have committed an even greater heresy than suggesting that the Buddha ate pork. They have shown that the *Viṣṇu Purāṇa*, where the story of the divine boar appears, was composed after the Buddha's death, "and that therefore, Buddha must have swallowed *bonâ fide* pork, and not merely an esoteric boar."[51] That is, textual research renders Sinnett's reading preposterous.

One might justifiably ask at this point what this obscure exchange between two late Victorians, one an aged Oxford Sanskritist and the other, at least in the view of some, an embittered spiritualist quack, could possibly have to do with the discourse of Buddhism and Science. At least one of the questions it raises is: Who is the Buddha, and who speaks in his name—the scholar or the devotee? (We will postpone for the moment the implications of the devotee's ethnicity in this case.)

The Theosophists were often at odds with European scholars of the Orient and disparaged them for their narrow-mindedness. In a letter to A. P. Sinnett from 1882, the Master Koot Hoomi writes: "Since those gentlemen—the Orientalists—presume to give to the world their *soi-disant* translations and commentaries on our sacred books, let the theosophists show the great ignorance of those 'world' pundits, by giving the public the right doctrines and explanations of what they would regard an absurd, fancy theory."[52] Elsewhere, Koot Hoomi mentions Rhys Davids by name[53] and also notes that "Karma and Nirvana are but two of the seven great MYSTERIES of Buddhist metaphysics; and but four of the seven are known to the best orientalists, and that very imperfectly."[54]

Despite his time in India, Sinnett would not have considered himself an orientalist; indeed he seems to take pride in not being counted in their number. Yet as editor of one of the most influential newspapers of the Raj (although not located in Delhi, Calcutta, or Bombay), he played an important role in the representation of India and Indians to a wide Anglophone readership. The exchange between Müller and Sinnett would thus seem to be yet another of the scores of orientalist squabbles that took place over the course of the nineteenth century, two Englishmen (although one of them only honorary) trading polite insults over who has the better understanding of Oriental wisdom, without the participation of any Orientals, or at least any real ones. For Sinnett's orientalism is heightened by the conceit that his knowledge derives from Aryan masters, communicating telepathically from deepest Tibet. But were they?

Readers of *The Mahatma Letters* have generally fallen into two camps. There are those who are members of the Theosophical Society, or sympathetic to it, who regard the letters as they are represented: communications from the masters Koot Hoomi and Morya to A. P. Sinnett, with Madame Blavatsky serving as postmistress and sometime scribe. And there are those, beginning shortly after their appearance, who have dismissed the letters as entirely the work of Blavatsky. Recently, however, the Theosophist K. Paul Johnson has sought to identify the numerous figures—Hindu, Buddhist, Masonic, Muslim, Parsi, Sikh; Indian, Egyptian, Persian, Sri Lankan, and at least one Tibetan—with whom Blavatsky and Olcott were associated during their travels. He speculates, for example, that the Master Koot Hoomi, whose full name was Koot Hoomi Lal Singh, was in fact Thakar Singh Sandhawalia, a founder of the Singh Sabha.[55]

Blavatsky and Olcott were active opponents of Christian missionaries in South Asia and allied themselves with various reform and independence movements in both India and Sri Lanka. The possibility that there were Indians involved somewhere along the chain of communication called *The Mahatma Letters* (as well as *Esoteric Buddhism* and *The Secret Doctrine*) raises a host of questions about orientalism and authority, perhaps the most outlandish of which is whether Madame Blavatsky's ventriloquism somehow allowed the subaltern to speak.

It is also important to note, however, that the allegiances that Blavatsky and Olcott forged with South Asians tended to be short-lived. Before they departed for India, Madame Blavatsky had told Olcott that Swami Dayananda Saraswati (1824–1883), founder of the Arya Samaj, was "an adept of the Himalayan Brotherhood inhabiting the Swami's body."[56] But by 1882 Olcott had concluded that the swami was just a swami, not an adept at all, who had expressed his vexation "to me—in very strong terms—that I should be helping the Ceylon Buddhists and the Bombay Parsis to know their religions better than heretofore, while, as he said, both were false religions."[57] Others, including such legendary figures as Vivekananda and Dharmapāla, after initially cordial relations with the Theosophists, would take exception to their claim that they could help Hindus and Buddhists "to know their religions better than heretofore" and would disavow any connection of their Hinduism and their Buddhism to Theosophy.

The modern would seem minimally to be that which is different from the ancient, and one of the marks of modernity, wherever it is located along a chronology, is the recognition of this difference. Other characteristics that might be added would be an emphasis on the mechanical over the organic, the individual over the group, differentiation over unity, the real over the transcendent, the existential over the metaphysical. Part of the continuing appeal of the Buddha is that he, at least since the mid-nineteenth century, has seemed so modern. Here, among the dizzying divinities of India, was a man with just one head and two arms. He had rejected the myths of ancient India that organized society into an oppressive caste system placing all power in the hands of priests who performed elaborate sacrifices in which they muttered unintelligible chants. He wrested that power from those priests and placed it in the hands of the individual, regardless of caste. He discovered the truth through his own efforts and then made it accessible to all, describing a universe in which there was no God, in which the transcendent unity of the Upaniṣads was replaced by the inexorable law of cause and effect, "the laying bare of the device," so to speak. His life had certainly been embellished by legends over the centuries, but European scholars, reading the most ancient scriptures, had been able to strip away these mythic accretions and metaphysical elaborations to reveal the rational man.

This is the Buddha created by the orientalists. This is the Buddha that Allen Ginsberg describes in an early poem: "He drags his bare feet out of a cave under a tree, / eyebrows grown long with weeping and hooknosed woe, / In ragged soft robes wearing a fine beard, unhappy hands / clasped to his naked breast—humility is beatness humility is beatness— / faltering into the bushes by a stream, all things inanimate / but his intelligence—stands upright there tho trembling."[58] This is the Buddha of Burnouf, who wrote in 1844:

> Indeed there are few beliefs that rest on so small a number of dogmas, and that also impose fewer sacrifices to common sense. I speak here in particular of the Buddhism that appears to me to be the most ancient, the human Buddhism, if I dare to call it that, which consists almost entirely in very simple rules of morality, and where it is enough to believe that the Buddha was a man who reached a degree of intelligence and of virtue that each must take as the exemplar for his life. I distinguish it intentionally from this other Buddhism of buddhas and bodhisattvas of contemplation, and above all from that of the Ādibuddha, where theological inventions rival the most complicated that modern Brahmanism has conceived. In this second age of Buddhism dogma develops, and morality, without disappearing entirely, is no longer the principal object of the religion.[59]

And this is the Buddha of Max Müller, a Buddha about whom Müller writes with real affection, as one who scorned secrecy, who taught for the sake of "the people at large, for the poor, the suffering, the ill-treated,"[60] who instructed his disciples to teach in the vernacular, who inveighed against the idea that his disciples should be guided by anything but the truth, for whom the greatest miracle was not telepathy but teaching, "by which an unbeliever is really converted into a believer, an unloving man into a loving man,"[61] who preached no dogma, who was born a man and died a man, after eating a meal of boar's flesh; "it has always seemed to me to speak very well for the veracity of his disciples that they should have stated this fact quite plainly."[62]

The Buddha of A. P. Sinnett is a very different being, as he explains in the chapter entitled "Buddha" in *Esoteric Buddhism*. It begins: "The historical Buddha, as known to the custodians of the Esoteric Doctrine, is a personage whose birth is not invested with the quaint marvels popular story has crowded round it. Nor was his progress to adeptship traced by the literal occurrence of supernatural struggles depicted in symbolic legend." Here Sinnett is in good company with the leading Oriental scholarship of the day. But he continues:

> On the other hand, the incarnation, which may outwardly be described as the birth of Buddha, is certainly not regarded by occult science as an event like any other birth, nor the spiritual development through which Buddha passed during his earth-life a mere process of intellectual evolution, like the mental history of any other philosopher. The mistake which ordinary European writers make in dealing with a problem of this sort, lies in their inclination to treat exoteric legend either as a record of a miracle about which no more need be said, or as pure myth, putting merely a fantastic decoration on a remarkable life. This, it is assumed, however remarkable, must have been lived according to the theories of Nature at present accepted by the nineteenth century.[63]

Sinnett thus clearly rejects the orientalist, and modernist, reading of the life of the Buddha, a reading that strips that life of its legendary character. Relying on a simplistic dichotomy between myth and history, miracle and fact, the orientalists are too quick to assume that an ordinary man, albeit a very good man, is to be found beneath the layers of legend, leaving only the life of a man lived in modern terms: teacher of virtue, social reformer, champion of the poor. He lived and died in ancient India, but he was a man to be admired even today. Substituting the word *Buddha* for the word *Christ* in his criticism of the professional scholars of Buddhism, Sinnett might just have easily have said:

> Finally, I declare that I am completely opposed to the error of the modernists who hold that there is nothing divine in sacred tradition; or what is far worse, say that there is, but in a pantheistic sense, with the result that there would remain nothing but this plain simple fact—one to be put on a par with the ordinary facts of history—the fact, namely, that a group of men by their own labor, skill, and talent have continued through subsequent ages a school begun by Christ and his apostles.

This quotation is from Pope Pius X's "Oath Against Modernism," delivered on September 1, 1910.

The Theosophists had no interest in this modern Buddha; they saw a different Buddha. But to see the Buddha in his true nature, a key is required: the act of interpretation. For hidden among the mass of fantastic elements that the orientalists dismiss as ancient superstition, the Theosophist discerns esoteric meanings that remain of vital importance to humanity.

Sinnett has a broader view of the Buddha than that of Müller and Burnouf. For Sinnett, the Buddha is just one of a series of adepts who have appeared over the course of the centuries. His next incarnation, occurring some sixty years after Gautama Buddha's death, was as Śaṃkara, the great Vedanta philosopher. Sinnett concedes that the uninitiated would place Śaṃkara's birth some thousand years after the death of the Buddha, and would also note Śaṃkara's rather virulent antipathy to Buddhism.[64] He reports that the Buddha appeared as Śaṃkara "to fill up some gaps and repair certain errors in his own previous teaching."[65] The Buddha had departed from the practice of earlier adepts by opening the path of the adepts to members of all castes. Although well intentioned, this led to a degradation of occult knowledge when it was transferred into unworthy hands. It thus became necessary thereafter "to take no candidates except from the class which, on the whole, by reason of its hereditary advantages, is likely to be the best nursery of fit candidates."[66] The Buddha did not reincarnate again until the fourteenth century, by which time the adept community had congregated in Tibet. Thus, the Buddha's next incarnation was as the Tibetan reformer Tsong kha pa.

Sinnett's pronouncements on the Buddha today seem fanciful, and objectionable. He takes that element of the Buddha's teaching that most appealed to Müller and Burnouf, and to so many Victorians—his commitment to teaching the dharma to members of all castes—and identifies it as an error; the occult knowledge should be revealed only to the most advanced hereditary group, the brahmans. Here Sinnett seems to echo the views of the Count de Gobineau (described in chapter 2) and the race theory Sinnett presumably learned from Madame Blavatsky.[67] But at the same time, his reading does something that is essential to the discourse of Buddhism and Science: in an act of cosmic colonialism, he extracts the Buddha from the conventional chronology of history and places him in a different chronology unknown but to the initiates, in which the Buddha is reborn as the great persecutor of Buddhism,

Śaṃkara, who is in turn reborn more than five hundred years later in Tibet, as Tsong kha pa. This is not unlike placing the Buddha in another lineage, one of such otherworldly geniuses as Galileo, Newton, Bohr, and Einstein, each of whom had his own interest in esoteric wisdom. In the nineteenth century, this act of interpretation was met by Asian teachers with bafflement or dismay (as in the case of Dayananda Saraswati). In each of these cases, the Buddha is separated from Buddhism and then placed in an imaginary lineage. Over the long history of Buddhism and Science, the Buddha would be repeatedly kidnapped and made to maintain simultaneously the authority of the ancient and the immediacy of the modern. Fortunately for the kidnappers, the Buddha had perfected the practice of patience aeons ago.

The Theosophical Society, and Madame Blavatsky, gained wide renown in the late nineteenth and early twentieth centuries, attracting interest, and members, throughout South Asia, Japan, Europe, and America, especially in the arts. In Great Britain, Theosophy captivated, among many others, several of the most famous Irish writers, including Shaw (very briefly), Yeats, and George William Russell (1867–1935), who published as "AE." In Russia, the pianist and composer Alexander Scriabin was a Theosophist, as were Kandinsky and Nicholas Roerich (1874–1947), a poet and artist who designed the costumes and sets for the scandalous first performance of Stravinsky's *Rite of Spring* in Paris in 1913. In the Netherlands, Mondrian was a member of the society. A writer whose artistic stature, for many Americans, equals or exceeds these figures was also a devoted Theosophist: L. Frank Baum, author of *The Wizard of Oz*.

As European interest in Theosophy waxed, Asian interest waned. The Theosophical Society continued to appropriate Buddhist doctrines. Madame Blavatsky had informed certain of her disciples that the purpose of Theosophy was to prepare the way for the coming buddha, Maitreya. Blavatsky's heir, the former British suffragette Annie Besant (1847–1933), selected a young Hindu boy in 1909 as the messiah Maitreya, the World Teacher of the Aquarian Age. (The boy, Jiddu Krishnamurti, renounced this status in 1929.) The American Theosophist Walter Y. Evans-Wentz discovered what he considered Theosophical doctrine in a Tibetan text that he would dub *The Tibetan Book of the Dead*.[68] But with few exceptions (D. T. Suzuki called Madame Blavatsky's *The Voice of the Silence* "the real Mahāyāna Buddhism"), Buddhist figures did not reciprocate the interest of the Theosophists. In 1905 the leading Buddhist monk in Sri Lanka withdrew his imprimatur from the fortieth edition of Olcott's *Buddhist Catechism*, declaring that seventeen of the answers were "opposed to the orthodox views of the Southern Church of Buddhism." Dharmapāla, the person who had been closest to Blavatsky and Olcott in their early efforts on behalf of Buddhism, was particularly emphatic in his repudiation of Theosophy. In 1906 he published an essay, "Can a Buddhist Be a Member of the Theosophical Society?" The short answer was no. Buddhism bore no historical relation to any other religion, and thus a "conscientious Buddhist who is well versed in Buddhist lore can no more sympathise with the principles of Theosophy than with the teachings of Christ, Muhammad, Krishna and Moses."[69]

Two decades later, he was more vociferous, writing in a letter of February 20, 1926, "Members of the Theosophical Society who follow [Charles W.] Leadbetter [1854–1934] and Mrs. Besant are against Buddhism. They follow Jesus and he they say is greater than our Lord Buddha. Leadbetter and Mrs. Besant steal everything from Buddhism and palm it off as their own and twindle the ignorant members of the T. S. in England."[70] Thus, Theosophy, which in an apparently ecumenical spirit had sought to unite the religions of the world through linking them back to an ahistoric and prehistoric wisdom, was now rejected by the Buddhists as a modern creation.

But having broken with Olcott and Theosophy, Dharmapāla took from them the view of the Buddha as not merely an ancient teacher to be admired across the mists of time, but a world historical figure of contemporary relevance. In his address to the World's Parliament of Religions in 1893 (long before his break with Theosophy) he had proclaimed, "Buddhism is a scientific religion, in as much as it earnestly enjoins that nothing whatever be accepted on faith. Buddha has said that nothing should be believed merely because it is said. Buddhism is tantamount to a knowledge of other sciences."[71] And he continued to proclaim that compatibility of Buddhism with Science—or perhaps more accurately, the compatibility of Science with Buddhism—over the decades, his words often bearing the trace of his Theosophical tutelage. In 1914: "With the expansion of knowledge Europeans may come to know more of evolution, of the laws of causation, of the changing nature of all phenomena, of the divisibility of matter, of the progressive nature of the animal and human consciousness, then will Buddhism meet with a sympathetic reaction."[72] In 1924: "Every new discovery in the domain of science helps for us to appreciate the sublime teachings of the Buddha Gautama."[73] In 1929: "Buddhism is for the scientifically cultured. The discoveries of modern science are a help to understand the sublime Dhamma. . . . To-day, the cultured races of Europe require a scientific psychology showing the greatness of the human consciousness. The sublime Doctrine of the Lord Buddha is a perfect science based on transcendental Wisdom. This Dhamma should be given freely to the European races."[74]

We thus see the claim of the Theosophist, that the most ancient is in fact the most modern, being claimed by the Buddhist, seeking to defend his religion against the attacks of various enemies, both foreign and domestic. Indeed, the claim for a compatibility of Buddhism and Science offers a kind of defense against modernity, even striking out against it, by proclaiming that an ancient Indian religion is a "science" and thus is modern.

But what became of the orientalists, reading their texts in the libraries of Europe and America, in search of the true Buddha? Following the instructions of Müller, scholars have copied, collated, and translated more and more Buddhist manuscripts in the Bodleian (and elsewhere). As they have done so, the Buddha of the nineteenth century—whether the social reformer of the Victorian scholars or the harbinger of spiritual evolution of the Theosophists—has given way to a Buddha who is less individual and more generic. We read, for example, that all of the buddhas who have come in the past and who will come in the future do a great many of the same things. They all sit cross-legged in their mother's womb; they are all born in the "middle country" of our continent of Jambudvīpa; immediately after birth they all take seven steps to the north; they all renounce the world after seeing the four sights (an old man, a sick man, a dead man, and a mendicant) and after the birth of a son; they all achieve enlightenment seated on a bed of grass; they stride first with their right foot when they walk; they never stoop to pass through a door; they all found a monastic order; they all can live for an aeon; they never die before their teaching is complete; and—Müller was right—they all die after eating meat. Four sites on the earth are identical for all buddhas: the place of enlightenment, the place of the first sermon, the place of descending from the Heaven of the Thirty-Three atop Mount Meru, and the place of the bed in Jetavana monastery.

Indeed, buddhas can differ from one another in only eight ways: in life span, height, caste (either brahman or *kṣatriya*), the conveyance in which they go forth from the world, the period of time spent in the practice of asceticism prior to their enlightenment, the kind of tree they sit under on the night of their enlightenment, the size of their seat there, and the circumference of their aura.[75] If we were to strip this traditional list of its mythological elements, the Buddha would be little more than a statue in Max Müller's hearth.

• • •

The Dalai Lama visited the University of Michigan on April 21–23, 1994. One of the events planned for his visit was a private seminar on the origins of the Mahāyāna with the faculty and graduate students of the Buddhist Studies program. I thought it would be interesting to discuss with him current Western scholarship on this topic, one that is both of great interest to the Dalai Lama and has been the subject of significant speculation in Buddhist studies in the last three decades. I wondered whether the Dalai Lama would be as interested in the findings of Western scholars of Buddhism as he has been in the findings of Western scientists. Among the Mahāyāna traditions of India, Tibet, China, Korea, and Japan, the accepted view is that the Mahāyāna sūtras were set forth by the historical Buddha Śākyamuni, but were kept hidden until some four centuries after the Buddha passed into nirvāṇa. Some of the sūtras were kept by the gods, others were kept by the nāgas (a kind of dragon or serpent) at the bottom of the sea, whence they were retrieved by Nāgārjuna. The authenticity of these sūtras as the word of the Buddha is a central issue for the Mahāyāna; to claim that they are inauthentic is to transgress one of the bodhisattva vows. When I mentioned to the Dalai Lama that our graduate students would be making a presentation to him on the origins of the Mahāyāna, he immediately asked whether they had supernormal powers, suggesting that only someone who had a clairvoyant knowledge of the past could know how the Mahāyāna began.

During the seminar, three students who were completing dissertations on Indian Buddhism made brief presentations to the Dalai Lama. They explained how nineteenth-century scholars of Buddhism had seen the Mahāyāna as a degeneration of the original teachings of the Buddha. Later scholars saw the Mahāyāna as a lay movement responding to the conservatism of the monastic establishment. After this perceived split, which occurred between the first century BCE and the first century CE, two branches of Buddhism, the Hīnayāna and Mahāyāna, developed along parallel but divergent courses. More recently, scholars have sought to look beyond the polemical Mahāyāna condemnations of the Hīnayāna and to consider archaeological, art-historical, and epigraphical evidence. This research suggests that the Mahāyāna did not begin as a single and self-conscious movement, but instead was a disparate collection of "cults of the book" centered on new sūtras composed around the beginning of the common era. These were not lay cults, but ones in which monks and nuns were full and active participants. The evidence suggests that so-called Mahāyāna and Hīnayāna monks often lived side by side within the same monasteries, following the same rules, engaging in many of the same practices, throughout the history of Buddhism in India. Indeed, the first epigraphic use of the term *Mahāyāna* occurs only in the fifth century CE, some five hundred years after the composition of the first Mahāyāna sūtras.

The Dalai Lama listened attentively to all of this, sometimes stopping and asking his translator to clarify a term or point. But at the end of the presentation he remained silent until I asked him for his thoughts on what the students had said. "It's something to know," he said in Tibetan, using a term, *shes bya* (literally, "object of knowledge"), that evokes a Buddhist aphorism: "Objects of knowledge are limitless." That is, there are infinite things that can be known, but some are more consequential than others; hence it is essential to consider carefully what is truly worth knowing. In this context, to call this information "something to know" suggested that it did not fall into the category of what is truly worth knowing. The Dalai Lama went on to say that he has a friend, a great lama, who, when giving a tantric initiation, saw each of the past masters of the lineage from previous centuries appear in the air along the ceiling of the temple. He was certain that his friend was telling the truth. He conceded that what the students had told him was interesting and that it would be good for Buddhists to have some knowledge of Western scholarship on Buddhism. However, in the end, he seemed to view Buddhist practice and Buddhist scholarship (at least of the Western variety) as ultimately irreconcilable. He told the students that if he accepted what they had told him, he would only be able to believe in the *rūpakāya*, the physical body of the Buddha that appears in the world to teach the dharma. He could not believe in the *saṃbhogakāya*, the body of enjoyment that appears to advanced bodhisattvas in the splendor of the pure lands, adorned with the thirty-two marks of a superman. And he could not believe in the *dharmakāya*, the Buddha's omniscient mind and its nature of emptiness. "If I believed what you told me," he said, "the Buddha would only be a nice person."

THE MEANING OF MEDITATION

A Tibetan monk sits down to meditate. Standing before him is a small altar, which holds a statue of the Buddha, a scripture (a block print wrapped in cloth), and a small reliquary. As offerings to these symbols of the body, speech, and mind of the Buddha, he has placed seven small brass bowls on the altar, each filled with water. A stick of incense burns in a bowl of uncooked rice. The monk is seated on a simple cushion, his legs in the lotus posture. He straightens his spine like a stack of coins and levels his shoulders, tucking in his chin so that his nose is aligned with his navel. He gently sets his teeth so they touch and places his tongue against the back of his upper teeth. Leaving his eyes open, he casts his gaze downward and breathes gently.

In the space in front of him the monk visualizes his own teacher as the buddha Vajradharma, in the form of a sixteen-year-old Indian prince. He is seated on a throne supported on the backs of eight great lions. The cushion of the throne is a lotus blossom; its eastern petal is white, its northern petal green, its western petal red, its southern petal yellow. The center of the lotus is green, and on it sits the buddha on cushions of the moon and the sun. Vajradharma's body is red, and his hands are crossed at his heart. His right hand holds a *vajra* (a ritual implement) and his left holds a bell. He sits in the lotus posture, and is adorned with ornaments made of human bone: a crown, earrings, a necklace, armlets and anklets, and an apron. His body is smeared with ashes from a cremation ground.

Inside the heart of Vajradharma, the monk imagines that there is a disc of the sun, upon which stands the buddha Cakrasamvara. His body is blue, with four faces (blue in the east, green in the north, red in the west, and yellow in the south) and twelve arms. One right hand holds a *vajra*, one left hand holds a bell, and together these two hands embrace his consort, who is red. Two hands hold the flayed skin of an elephant. His other right hands hold a drum, an axe, a cleaver, and a trident. His other left hands hold a *khatvaṅga* (an eight-sided staff surmounted by three heads and a *vajra*), a skull cup, a noose, and the head of the god Brahmā.

On the eastern petal of the lotus upon which Vajradharma sits (the petal directly in front of the monk) stands Vajrayoginī, surrounded by the deities of the four classes of tantras. On the southern petal (the petal to the left of the monk) sits the buddha Śākyamuni, surrounded by the buddhas of the ten directions. On the western petal (the petal to the right of the monk) are all the scriptures of the sūtras and tantras, in the form of books made of light. On the petal to the north sits Mañjuśrī, the bodhisattva of wisdom, surrounded by bodhisattvas and arhats. Around the entire assembly are the wrathful deities who protect the dharma.

The monk imagines that he is not alone in the presence of this august assembly but surrounded by all the beings in the universe, beginning with his own parents. He considers how rare it is to be reborn as a human who has the opportunity to benefit from the Buddha's teachings. He realizes that death can occur at any instant and that, as a result of his past karma, he could be reborn in one of the unfortunate realms of the cycle of rebirth: as an animal, ghost, or hell being. He thus determines to make every effort to achieve liberation from rebirth now. Visualizing the assembly of buddhas, bodhisattvas, and deities before him, the monk then silently recites a prayer three times, going for refuge to the teachers and to the three jewels: the Buddha, the dharma, and the saṅgha. He imagines that all the beings in the universe join him in this prayer. In response, each member of the divine assembly sends forth a stream of light and nectar in five colors (blue, red, yellow, green, and white) that enters into the monk and all beings, purifying them of their negative karma.

The monk next turns his thoughts to the plight of others, the various gods, demigods, humans, animals, ghosts, and hell beings who populate the universe. Each of them is subject to karma and rebirth, each undergoes unlimited forms of sufferings as they wander endlessly through the realms of saṃsāra. Because this cycle of rebirth has no beginning, the monk understands that he has been in every possible relationship with each being in the universe and that each of these beings has therefore been his human mother in a past life. Remembering their kindness, he is overwhelmed with feelings of pity for their sad plight and determines to free them from suffering, vowing to become a buddha himself so that he might lead all beings to liberation.

He cultivates love, wishing that all beings would have happiness; he cultivates compassion, wishing that all beings would be free from suffering; he cultivates joy, wishing that all beings would remain happy forever; and he cultivates equanimity, wishing that all beings would remain equally happy. In the presence of this esteemed assembly, he vows to achieve buddhahood for the sake of all beings and to never forsake that vow, even if it costs him his life.

The monk maintains three categories of vows: monastic, bodhisattva, and tantric. He mentally reviews these categories. He begins by promising not to transgress any of the 253 vows of a fully ordained Buddhist monk. He then mentally promises not to commit the eighteen root infractions of the bodhisattva vows: (1) to praise oneself and slander others out of attachment to profit or fame; (2) to withhold one's wealth or the doctrine, out of miserliness, from those who suffer without protection; (3) to become enraged and condemn another, without listening to his or her apology; (4) to abandon the Mahāyāna and teach a facsimile of the excellent doctrine; (5) to steal the wealth of the three jewels; (6) to abandon the excellent doctrine; (7) to steal the saffron robes of a monk, and beat, imprison, and expel a monk from the life of renunciation, even if he has broken the ethical code; (8) to commit the five deeds of immediate retribution: to kill one's father, to kill one's mother, to kill an arhat, to wound a buddha, or to cause dissent in the saṅgha; (9) to hold wrong views; (10) to destroy cities; (11) to discuss emptiness with sentient beings whose minds have not been trained; (12) to turn someone away from seeking buddhahood; (13) to cause someone to abandon completely the monastic vows in order to practice the Mahāyāna; (14) to believe that such afflictions as desire cannot be abandoned by the vehicle of *śrāvaka*s, and to cause others to believe it; (15) to claim falsely, "I have withstood the profound emptiness"; (16) to impose fines on renunciants; to take donors and gifts away from the three jewels; (17) to cause meditators to give up the practice of serenity (*śamatha*); to take the resources of those in meditation retreat and give those resources to reciters of texts; (18) to abandon the two types of the aspiration to enlightenment (*bodhicitta*), the aspirational and the practical.

Next he promises not to commit the twenty-eight infractions of the tantric vows: (1) to use an unqualified consort; (2) to engage in union inappropriately; (3) failing to keep secrets from the unripened; (4) to argue in an assembly; (5) to teach the faithful a doctrine different from that which they requested; (6) to stay seven days with a *śrāvaka*; (7) to falsely claim to be a yogin; (8) to teach the doctrine to the faithless; (9) to display secrets to the unripened without preparing them well; (10) to enter a maṇḍala without proper purification; (11) to transgress the monastic and bodhisattva vows without purpose; (12) failing to promise to restrain oneself from the root infractions and the gross infractions six times a day; (13) to go against the three ethical codes; (14) failing to go for refuge six times a day; (15) failing to rely on the pledge of the *mudrā*, *vajra*, and bell; (16) failing to give the gifts of requisites, fearlessness, the doctrine, and love six times a day; (17) failing to create the wish to uphold the excellent doctrines of the three outer vehicles and secret vehicle; (18) failing to strive to keep the vows and to make offerings six times a day; (19) failing to give up what is unsuitable, except in order to tame sentient beings; (20) failing to make offerings to yogins; (21) failing to strive to practice the ten virtues; (22) to desire the Hīnayāna; (23) to turn one's back on the welfare of sentient beings; (24) to abandon saṃsāra; (25) to be constantly attached to nirvāṇa; (26) to scorn gods, demigods, and secret deities; (27) to walk over or eat in the presence of *mudrā*s, conveyances, weapons, and hand implements of deities; (28) to step on the shadows of ancient deities.

The monk then prays to receive the blessings of the three jewels—the Buddha, the dharma, and the saṅgha—assembled before him. In response, the saṅgha jewel transforms into white light and enters the monk's body through the crown of his head, blessing his body. The dharma jewel (including all of the scriptures) transforms into red light and enters the monk's throat, blessing his speech. The Buddha jewel (including all of the buddhas) transforms into blue light and enters the monk's heart, blessing his mind.

Next the monk visualizes a red letter *baṃ* in the center of his heart. Above it is a tiny crescent moon and a flaming sphere. He imagines the letter slowly expanding in size. As it does so, his body dissolves into red light, beginning at the heart and moving outward. The letter *baṃ* continues to expand until it pervades the entire universe, with all beings and the environments they inhabit transformed into an immense letter *baṃ*. The letter then slowly begins to contract, leaving only empty space in its wake, until the letter has become very tiny. Now the letter itself begins to dissolve into space from below, leaving only the crescent moon and the flaming sphere. The crescent moon dissolves into the sphere, the sphere into the flame. The flame then disappears from below, leaving only the clear light of emptiness.

After resting in emptiness, the monk imagines that his mind emerges from this emptiness in the form of the letter *baṃ*. Two letters then appear in space, transforming themselves into two interlocking inverted three-sided pyramids. Viewed from above, their flat upturned base is in the shape of a hexagram. A white letter *āḥ* appears in the center of the hexagram and turns into a white moon disc. Standing upright around the edge of the disc are the letters of the thirty-two-syllable mantra, red in color. The letter *baṃ* descends through space and stands in the center of the moon disc. Rays of light then radiate from the letters throughout all of space, purifying all beings and the environments they inhabit. Everything again melts into light and dissolves into the letter *baṃ*, which is immediately transformed into the maṇḍala of Vajrayoginī.

The maṇḍala, in the shape of a circle, is surrounded by a fence and enclosed under a canopy of five-pronged flaming blue *vajra*s, protecting the interior from all dangers and intruders. Immediately within the fence lie eight charnel grounds, in the east, north, west, south, southeast, southwest, northwest, and northeast. Each charnel ground has a tree, a guardian, a protector, a lake, a serpent deity (nāga), a cloud, a fire, and a stūpa. In addition, in each charnel ground there are crows, owls, eagles, jackals, snakes, various carnivorous spirits, and yogins and yoginīs. All of this is visualized by the monk.

Within the circle of charnel grounds is a divine palace in the shape of the inverted interlocking pyramids. In the center is an eight-petalled lotus. The petals in the east, south, west, and north are red. The southeast petal is yellow, the southwest petal green, the northwest petal yellow, and the northeast petal black. The center of the lotus is green, surmounted by a sun disc.

The monk visualizes himself as Vajrayoginī at the center of the sun disc. Standing upright, her right foot rests on the chest of the red goddess Kalarati. Her left foot rests on the back of the black god Bhairava. Her body is brilliant red, the color of fire. She has one face and three eyes, each looking upward. In her right hand she holds a cleaver. In the bend of her left arm she holds a *khatvaṅga*, at the top of which are three human heads—blue, red, and white—and a five-pronged *vajra*. Hanging from the staff are a drum, a bell, and three banners. In her left hand she holds aloft a skull cup filled with blood. Her black hair hangs down to her waist. She is naked, adorned with a crown of five human skulls; a necklace of fifty human skulls; and earrings, armlets, anklets, and an apron, all made from human bone.

The monk next turns to the visualization of the body maṇḍala, in which thirty-seven parts of his body (visualized as Vajaryoginī) are transformed into thirty-seven deities. The thirty-seven places comprise the twenty-four inner places of the body, the eight channels of the sense organs, the four channels of the heart wheel (*cakra*), and the indestructible drop. The twenty-four inner places are the hairline, the crown of the head, the right ear, the back of the neck, the left ear, the point between the eyebrows, the eyes, the shoulders, the armpits, the breasts, the navel, the tip of the nose, the mouth, the throat, the heart, the testicles, the tip of the genitals, the anus, the thighs, the calves, the eight fingers and toes, the tops of the feet, the thumbs and big toes, and the knees.

The remaining thirteen places are connected with the tantric system of physiology, in which a network of seventy-two thousand channels radiate throughout the body. Among all these channels, the most important is the central channel. As straight and as thick as an arrow, it is red in color, running from the genitals upward to the crown of the head, then curving down to end in the space between the eyes. Located within the central channel are the red drop and the white drop, which derive from the mother and father, respectively. Running parallel to the central channel are the red right and the white left channels, connected to the central channel at the navel center and ending at the right and left nostrils. These two channels wrap around the central channel at several points, creating constrictions that prevent wind from moving through the central channel. At these points of constriction, networks of smaller channels radiate throughout the body in increasingly smaller channels. These points are called wheels (*cakra*s). There are seven of these wheels: at the forehead, the crown of the head, the throat, the heart, the navel, the base of the spine, and the opening of the sexual organ.

The eight channels of the sense organs connect the tongue, the navel, the genitals, the anus, the point between the eyebrows, the ears, the eyes, and the nostrils to the central channel at the point of the heart *cakra*. The four channels of the heart wheel are the four main channels radiating from the heart *cakra*. The indestructible drop is a small sphere, the size of a small pea, located at the center of the heart *cakra*. White on top and red at the bottom, it encases the mind in clear light.

In the practice of the body maṇḍala, these thirty-seven physical elements are transformed into thirty-seven letters and then into thirty-seven deities. At the center of the heart *cakra*, the monk visualizes the inverted interlocking pyramids surmounted by a tiny horizontal moon disc. Thirty-two channels, short in length and transparent, are connected to the edge of the moon disc, rising vertically, each slightly thicker than a needle. Within this circle of channels are the four channels of the heart *cakra*, located in the cardinal directions. In the center of the moon disc stands the indestructible drop, white on top, red on the bottom, yet translucent. Each of these thirty-seven elements then transforms into a letter. The thirty-two channels around the periphery are transformed into the letters of the thirty-two-syllable mantra. The four channels are transformed into a green *ya* in the north, a red *ra* in the west, a yellow *la* in the south, and a white *va* in the east. The indestructible drop at the center of the moon disc turns into the letter *aḥ*, white at the top and red at the bottom. The letters are next transformed into goddesses, in the sequence of the recitation of the syllables of the mantra. First the thirty-two letters of the mantra turn into thirty-two different goddesses. The four letters in the cardinal directions then transform into the green goddess Lāmā in the north, the red goddess Khaṇḍārohā in the west, the yellow goddess Rūpinī in the south, and the white goddess Ḍākinī in the east. The letter in the center of the moon disc transforms into Vajrayoginī herself.

Through years of practice, the monk is able to visualize the entire maṇḍala, including each of the deities and their individual attributes, in minute detail. He is able to shrink the entire maṇḍala to the size of a mustard seed and visualize the maṇḍala precisely, down to the color of each goddess's eyes, with uninterrupted, single-pointed concentration for four hours.

Visualizing the entire maṇḍala, with himself in the form of Vajrayoginī, the monk recites the mantra *phaiṃ* and forms the hand gesture (*mudrā*) called "blazing," while imagining that rays of light burst forth from the letter *baṃ* at the heart of Vajrayoginī. Radiating from the place between her eyebrows, the rays of light travel throughout space, touching all the buddhas of the ten directions, who thereby take on the form of Vajrayoginī. The rays of light also reach to the heaven called Akaniṣṭha ("unsurpassed"), where the actual Vajrayoginī resides. All the buddhas in the form of Vajrayoginī converge into a single Vajrayoginī. With the monk's recitation of the mantra *jaḥ hūṃ baṃ hoḥ*, with each syllable accompanied by a different *mudrā*, the actual Vajrayoginī descends from her heaven to stand in space above the head of the monk (still visualizing himself as Vajrayoginī). She dissolves into him, she is fused with him, and he becomes the actual Vajrayoginī. Next, the monk performs another *mudrā* and recites the mantra *oṃ yoga śuddhāḥ sarvadharmāḥ yoga śuddho 'haṃ* (which means, in Sanskrit, "All phenomena are pure yoga. Pure yoga am I").

In order to maintain the absorbed deities within and to protect against the intrusion of negative forces from without, now the monk engages in a practice called "donning armor." While visualizing himself as Vajrayoginī, he also visualizes six goddesses at various points of the body, between the skin and the flesh, in the form of pairs of letters. Two red letters stand upright at the navel, two blue letters stand upright at the heart, two white letters stand upright just below the chin, two yellow letters stand upright at the forehead, two green letters lie flat at the crown of the head, two smoke-colored letters stand upright at the shoulders, wrists, hips, and ankles. While still visualizing Vajrayoginī, brilliant red in color, the monk imagines that each of these pairs of letters radiates its own color, encasing Vajrayoginī's body in a protective aura: the light around the crown of the head is green, the light from the forehead to the chin is yellow, the light from the chin to the heart is white, the light from the heart to the navel is blue, the light from the navel to the hips is red. The arms and hands, legs and feet are surrounded by smoke-colored light.

Red rays of light once again radiate from the red letter *baṃ* and the thirty-two syllables of the surrounding mantra at the heart of Vajrayoginī. These pervade the universe, purifying the negative karma of sentient beings and making offerings to all of the buddhas throughout space. The rays of light return in the form of blessings and dissolve into the letter *baṃ* and the surrounding syllables. The monk next repeats the thirty-two-syllable mantra.

Once again the monk, still as Vajrayoginī, visualizes rays of light emanating from the letter *baṃ* and the syllables of the mantra, extending into the three realms of the universe: the Formless Realm, the Realm of Form, and the Realm of Desire. As a result, the Formless Realm turns into rays of blue light that dissolve into the upper part of Vajrayoginī's body. The Realm of Form turns into rays of red light that dissolve into the middle part of Vajrayoginī's body. The Realm of Desire turns into rays of white light that dissolve into the lower part of Vajrayoginī's body. Now the body of Vajrayoginī itself begins to dissolve from above and below, like breath on a mirror, leaving only the interlocking pyramids, the moon disc, and the deities standing upright upon it, with Vajrayoginī at the center of the maṇḍala. The pyramids gradually dissolve into the moon disc, and the moon disc dissolves into the circle of the thirty-two goddesses. The thirty-two goddesses dissolve into the four goddesses in the cardinal directions. The four goddesses dissolve into Vajrayoginī at the center. Vajrayoginī then begins to dissolve into light from below and above, leaving only the interlocking pyramids at her heart, surmounted by a moon disc, where the letters of the thirty-two-syllable mantra stand upright around the letter *baṃ*. The interlocking pyramids dissolve into the moon disc, which dissolves in the thirty-two-syllable mantra, which dissolves into the letter *baṃ*. The letter *baṃ* now stands alone, surmounted by a small upturned crescent moon and a tiny flaming sphere. The letter begins to dissolve from below, leaving only the crescent moon and the flaming sphere. The crescent moon dissolves into the sphere, the sphere into the flame. The flame then disappears from below, leaving only the clear light of emptiness. The mind of the monk rests in this emptiness.

From within this emptiness, the monk spontaneously emerges, again in the form of Vajrayoginī, again with six goddesses at various points of her body, between the skin and the flesh, in the form of pairs of letters. Two red letters stand upright at the navel, two blue letters stand upright at the heart, two white letters stand upright just below the chin, two yellow letters stand upright at the forehead, two green letters lie flat at the crown of the head, and two smoke-colored letters stand upright at the shoulders, wrists, hips, and ankles.

Now the monk turns to the second stage of the practice, called the stage of completion, where he will seek to generate special forms of bliss within his body and then use the mind of bliss to realize emptiness, the profound nature of reality. To do this, he maintains the visualization described above, but focuses especially on himself in the form of Vajrayoginī, red in color, with a body made of light. Here he will subdue the winds (*prāṇa*) or subtle energies that serve as the vehicles for consciousness by withdrawing the winds from the network of seventy-two thousand channels that run throughout the body and causing them to enter the central channel. In order to do this, he focuses first on the navel center. There he visualizes the letter *raṃ*, the seed syllable of fire. Like the letter *baṃ*, it is surmounted by a small upturned crescent moon and a tiny flaming sphere. The letter is tiny, no larger than a sunflower seed. It is red in color and radiates warmth and light. The monk contracts the lower muscles of his torso and draws in his stomach, imagining that the winds that course through the lower part of the body are entering the central channel and gathering at the navel center, just below the letter *raṃ*. He next inhales and swallows, imagining that all the winds that course through the upper part of the body are entering the central channel, just above the letter *raṃ*.

The monk then focuses his mind on the wheel located at his navel and specifically on the tiny red *raṃ* within the central channel. As he concentrates on the letter, it begins to glow like heated metal and the tiny flame above the letter burns with bright light and emits intense heat. That heat rises slowly up the central channel, first to the wheel at the heart, then to the wheel at the throat, and finally to the wheel at the crown of the head. When it reaches the crown, the heat of the inner fire begins to melt the white drop located there, causing that drop to begin to melt. As it melts it begins to descend through the central channel, first to the wheel at the throat, then to the wheel at the heart, and then to the wheel at the navel, ultimately the wheel at the end of the central channel. As the drop slowly descends through each wheel, a different bliss is experienced.[1]

As the monk contracts the muscles of his lower body, his concentration is broken by the discomfort of the rectal thermometer and the electrodes attached to his body. He is performing his meditation not in a cave in northern India, but in a laboratory in New England, where a group of neuroscientists are seeking to determine whether his body temperature increases when he imagines the flaming letter in the central channel. They determine that it does.[2]

• • •

Over the past twenty-five years, the effects of Buddhist meditation have begun to be measured by neurologists, adding a new dimension to the Buddhism and Science discourse.[3] Rather than pointing to affinities between particular Buddhist doctrines and particular scientific theories, research on meditation has sought to calculate the physiological and neurological effects of Buddhist meditation. Such research would seem to introduce a welcome empirical element to the discourse.

This new dimension, often referred to as Buddhism and Psychology or, more narrowly, Buddhism and Cognitive Science, deserves a broader treatment than can be provided here. The assertions being made in this domain are qualitatively different from declaring, for example, that the Buddha understood the theory of relativity. The claim here is that Buddhist meditation works. However, in order to understand the laboratory findings, such a claim requires that one first identify what is *Buddhist* about this meditation, describe what the term *meditation* encompasses in this case, and explain what *works* means, especially in the context of the exalted goals that have traditionally been ascribed to Buddhist practice. Although these goals are numerous and variously articulated across the tradition, it can be said that their ultimate aim is not self-help but a radical reorientation toward the world—and in many articulations, a liberation from it—either for oneself or for all beings.

It is certainly the case that in the sūtras, the Buddha offers a range of advice to laypeople on how to live a happy and prosperous life; what might be termed "self-help." For example, in a work called the *Sigālaka Sutta*, the Buddha explains that there are six negative consequences of wandering the streets at odd hours: (1) one is unprotected, (2) one's wife and children are unprotected, (3) one's property is unprotected, (4) one is suspected of crimes, (5) one is the subject of rumors, and (6) one encounters all manner of unpleasantness.[4] There are countless stories of bodhisattvas giving myriad gifts, including their own body parts, to relieve the physical and mental suffering of others. Meritorious deeds, especially charity, are prescribed for a happy rebirth in the future. Instruction in meditation is rarely offered to the laity as a means to happiness in this life; it is offered to those seeking the goal of enlightenment, however that might be defined.

Indeed, it is useful to recall that the vast majority of Buddhists over the course of Asian history have not practiced meditation. It has traditionally been regarded as something that monks do, indeed, that only some monks do; the monastic codes make repeated reference to the needs of meditating monks, suggesting that they represented a group of specialists within the monastic order. Indian texts speak of monks who devote their efforts to one of three activities: study (which entailed the recitation of texts), meditation, and service to the monastery. The famous sixth-century Chinese text *Gaoseng zhuan* (Lives of Eminent Monks) lists ten categories of monastic vocation: translators, exegetes, theurgists, meditators, disciplinarians, self-immolators, cantors, promoters of good works, hymnodists, and sermonizers, and these categories pertain only to the eminent. In the Theravāda cultures of Sri Lanka and Southeast Asia, there has been a long tradition of dividing monastic practice into two categories: the vocation of texts and the vocation of meditation. In commentaries dating from as early as the fifth century, a preference was expressed for the former. Able monks were expected to devote themselves to study, with meditation regarded as the vocation of those who were somehow less able, especially men who were ordained late in life. And there are major forms of Buddhism, most notably the Pure Land traditions, in which the practice of meditation does not play a central role at any stage of the path, except perhaps at an early stage, where the conviction that meditation is futile might lead to the wish for birth in Amitābha Buddha's pure land.

And for those Buddhists who have practiced meditation over the centuries, there have been a vast array of forms, beyond the Zen meditation and mindfulness that are best known in the West. One reason for opening this chapter with a tantric visualization was to give the reader some sense of how complicated, and baroque, meditation practice can be. Vajrayoginī meditation is performed in all sects of Tibetan Buddhism, where the stage of generation and the stage of completion are deemed essential to attain buddhahood. The description provided above was in fact minimal, omitting, for example, the names, the colors, the postures, and the accoutrements of the thirty-two goddesses who stand in a circle at the edge of the moon disc, each of which the meditator must be able to visualize in precise detail. None of the symbolism of the deities and the maṇḍala was described. Nor was the deeper purpose of the practice explained. Gendun Chopel, one of the two Tibetans from chapter 3, described the purpose of that practice in a controversial essay on Buddhist thought published in 1952: "Thus, in general, in the Mahāyāna and also, especially, in all of the Vajrayāna, from the point at which it is suitable to view the lama as a buddha through to meditating on yourself as Vajradhara and believing that you have a fully established body maṇḍala—these are only for the purpose of turning this present valid knowledge upside down."[5] Tantric meditation is practiced daily by the Dalai Lama.

Even for *vipassana*, the "mindfulness" practice widely studied by cognitive scientists, the goal is something far beyond anything easily measured on an fMRI. In the text that is the *locus classicus* for the practice of mindfulness (and hence for the modern *vipassana* movement), the Buddha declares:

> Monks, this is the direct path for the purification of being, for the surmounting of sorrow and lamentation, for the disappearance of pain and grief, for the attainment of the true way, for the realization of Nibbāna—namely, the four foundations of mindfulness. . . . If anyone should develop these four foundations of mindfulness in such a way for seven days, one of two fruits could be expected for him: either final knowledge here and now, or if there is a trace of clinging left, non-return [to rebirth in the Realm of Desire].[6]

Research on meditation in the realm of cognitive science has taken two major forms. In the first, scientists seek to evaluate the efficacy (variously defined) of a limited number of types of meditation. Here, despite previous research on Transcendental Meditation (TM), the recent focus of neurological investigation has been on "Buddhist meditation." Yet most forms of meditation being studied are not the elaborate visualizations described earlier but practices that, from the Buddhist perspective, are shared by non-Buddhist traditions and that result in deep states of concentration leading to heavenly rebirth—but do not, in and of themselves, lead to nirvāṇa. Some researchers have explained that Buddhism has been selected because its traditions of meditation are more developed than those of other religions.[7] However, more salient factors would seem to include the religious predilections of the researchers involved and the undeniable charisma of the Dalai Lama. Thus, regarding the phrase "Buddhist meditation," one might ask: what constitutes a particular practice as "Buddhist," as distinct from an element of a larger yogic tradition found in a wide range of traditions, including Hinduism, Jainism, Sikhism, and Sufism, or contemplative practice in Daoism, Judaism, or Christianity? The effects of meditation on neurological function are clearly a promising arena for research. However, the focus on Buddhism for this research appears as yet another manifestation of the West's fascination with Buddhism—ever ancient, ever modern—as the most appropriate partner of Science.

The second form of neurological research involves using highly trained meditators as informants in the laboratory, interviewing them about their experiences, experiences that can also be measured using brain imaging. Here scientists are exploring possible correlations between first-person experience and more standard scientific data. Again, this arena of research holds great promise. But it raises important issues about translation: cross-linguistic (few master meditators speak fluent English), cross-cultural (in the Tibetan tradition, it is considered inappropriate to speak of one's meditative experiences, except with one's teacher), and cross-disciplinary (the monastery and the laboratory occupy quite different worlds). An advanced practitioner of Buddhist meditation would describe his or her experiences in the often highly technical vocabulary of Buddhist doctrine. How, then, are correlations to be made? Furthermore, this form of research is predicated on the assumption—one that has long lain at the heart of many of the claims concerning Buddhism and Science—that Buddhist doctrine is the product of Buddhist insight, that the chief constituents of Buddhist philosophy are the articulations of someone's (usually the Buddha's) experience in meditation.[8] However, it can be equally argued that it is not meditation that produced doctrine but doctrine that produces meditation. These are some of the issues that might be addressed as research on Buddhist meditation proceeds.

CONCLUSION

Measuring the Aura

In the end, how should we understand the pairing of Buddhism and Science? Many forms of Buddhism speak of the decline of the dharma, the process whereby Buddhism slowly disappears in the centuries after the Buddha's passage into nirvāṇa. There are numerous prophecies in the Buddhist scriptures about how long his dharma will last, with the period of the duration ranging from as short as 500 years to as long as 12,000 years, with the figures of 700, 1,000, 1,500, 2,000, 2,500, and 5,000 years also found.[1] In the *Manorathapūraṇī* (The Fulfillment of Wishes) by the great fifth-century Theravāda scholar Buddhaghosa, a chronology of 5,000 years is provided, in which the dharma gradually disappears over five periods of one thousand years each. During the first millennium after the passing of the Buddha, there will be a disappearance of the attainments, at the end of which no disciple will have the capacity to attain the initial and lowest of the ranks of the enlightened, called "stream-enterer." During the second millennium, there will be a disappearance of the method, at the end of which no disciple will be able to attain meditative states or maintain the precepts. During the third millennium, there will be a disappearance of learning, at the end of which all books of the canon will be lost. During the fourth millennium, there will be a disappearance of the signs of monastic life, at the end of which all monks will stop wearing saffron robes and will return to lay life. During the fifth and final millennium, there will be a disappearance of the relics, at the end of which the relics of the Buddha will reassemble and, after being worshipped by the gods, will burst into flame. If the practice of the four foundations of mindfulness is reduced to stress reduction[2] and the visualization of Vajrayoginī is understood as a technique for raising one's body temperature, where do we stand in the process of the disappearance of the dharma?

To ask such a question is to imply that change is loss. When change is uncontrollable, it has long been regarded in Buddhism as a form of suffering. Yet Buddhism has undergone remarkable changes over its history, up to the present. And in the face of change, Buddhist thinkers have struggled to control the meaning of the term *Buddhism* (and its often rough equivalents in Asian languages) and the meaning of the term *Buddhist*. As proponents of Buddhism have continued over the decades to expand what is encompassed by the term in order to accommodate various forms of science, it is perhaps useful at this point to pause briefly to consider the question of what makes a person, or an idea, "Buddhist."

A Buddhist is traditionally defined as a person who regards the three jewels—the Buddha, the dharma, and the saṅgha—as the true source of refuge from the sufferings of the world. The Buddha, the dharma, and the saṅgha (generally translated as "community") are called jewels because they are difficult to encounter, and if encountered are of great value. But there are detailed commentaries on what constitutes each of the three jewels, with a range of opinion set forth concerning each of them.

A doctrine or assertion (in contrast to a person) is defined as Buddhist if it is marked by, that is, consistent with, what are known as the four seals: (1) that all products are impermanent, (2) that all contaminated things are miserable, (3) that all phenomena are empty and selfless, and (4) that nirvāṇa is peace. But these criteria are composed of the technical terms of Buddhist philosophy, each of which is again subject to considerable commentary, even among the Indian schools of Buddhist thought that flourished before Buddhism spread across Asia. And since being introduced to the West, the question of the meaning of the last element in the list, nirvāṇa, has generated long and often contentious comment. On the one hand, it has been labeled "annihilation," in light of standard

Buddhist views that nirvāṇa is the cessation of mind and body. On the other, it has been seen as simply an exalted state of mind. As we read in an essay of 1885, "But here not even Gotama Buddha has expressed the aspiration so plainly and nobly as a modern idealist. George Eliot's hope to live 'in thoughts sublime that pierce the night like stars, and with their mild persistence urge man's search to vaster issues,' is the most perfect, as it is the most poetical, description of Nirvana."[3]

As the Buddhist tradition developed in India after the Buddha's death, new doctrines inevitably arose, and means had to be found to deal with them. One of the more famous of these is the so-called four great authorities (*mahāpadeśa*), a test for determining whether the words that a monk reports to have heard from one of four sources of authority are the teaching of the Buddha. The four are the words of (1) the Buddha, (2) a community (*saṅgha*) of elders, (3) a smaller group of learned elders, and (4) a single learned monk. It is explained that when a monk claims to have heard a teaching directly from one of these four sources, the community of monks should determine whether it is the word of the Buddha by seeing whether it (1) accords with the recognized teachings of the Buddha (the sūtras) and (2) is in agreement with the monastic code (the vinaya). If it passes this test, it is to be accepted; if it does not, it is to be rejected.[4] This method appears to be the product of a community simultaneously lamenting the loss of teachings already forgotten and hence seeking to discover and preserve whatever might still be remembered by someone, while being wary of the introduction of innovation.

But innovation inevitably occurred, and techniques were developed to accommodate it. One of these was the division of scriptures into the definitive (*nītārtha*) and the provisional (*neyārtha*), encountered briefly in chapter 1. Although again variously interpreted, the pair was often used to distinguish those things that the Buddha said which he also believed from those things that he said in deference to the limited understanding of his audience.[5] This strategy would seem to encourage rather than discourage innovation, for it allowed an old teaching, once regarded as authoritative, to be superseded without being rejected. Indeed, nothing could be rejected, because all was the word of the enlightened Buddha. The Polish scholar Stanislaw Schayer (1899–1941), noting that a text judged definitive by one Buddhist school was classified as provisional by another, speculated that "the distinction between *nītārtha* and *neyārtha* used by Buddhist exegetes is a merely clerical stratagem and that all the texts preserved in Buddhist records originally possessed only one direct and literal meaning."[6]

Innovation recurred with the rise of Buddhist tantra, when all manner of sex and violence was attributed to the buddhas and bodhisattvas; some of that imagery appears in the meditation practice described at the beginning of the previous chapter. When such texts were first read by European orientalists, they were horrified. Only later did they read the commentaries, where the sex and the violence are often declared "symbolic."

With the rise of Buddhist modernism in the twentieth century, a new group of texts was added to the canon, texts by D. T. Suzuki, Alan Watts, and Fritjof Capra, works that both generated interest in Buddhism and defined it, providing yet another standard by which that which came before becomes "provisional." Such authors might be dismissed as amateurs in some quarters. But the situation becomes more complicated when a figure who speaks with traditional authority for a significant segment of the Buddhist world, a figure such as the Dalai Lama, propounds the modernist view of Buddhism and Science (at least in his English-language teachings), in a sense canonizing it.

The history of science is often represented as a process of replacement, as a new model, often incommensurable with what had been accepted previously, displaces its predecessor. The old models remain, but only as moments in the history of science. Thus, the heliocentric universe of Copernicus replaced the geocentric universe of Ptolemy; Lavoisier's theory of chemical reaction replaced alchemical transmutation; the theory of evolution replaced biblical creationism; quantum mechanics replaced (or at least required the revision of) Newtonian mechanics.

Buddhism, without a synod or a pope to declare what is orthodox and what is heterodox, became a tradition in which nothing is discarded, although something may be forgotten. New texts continued to be added, each claiming to be the word of the Buddha, with what was once definitive now being deemed provisional. All accretions were somehow accommodated. Yet the origins remain sufficiently occluded to make it possible to ask: is there only accretion?

Long ago, in ancient India, an innocuous statement carved in stone at the order of the emperor Aśoka in the third century BCE was reversed. Aśoka had proclaimed, "All that the blessed Buddha has spoken is well spoken." We find in a later sūtra, "All which is well-spoken, Maitreya, is spoken by the Buddha." This reversal would seem to remove all restrictions on admission to the category of the word of the Buddha. But the sūtra, perhaps not surprisingly, glosses the term, and qualifies what it means to be "well-spoken" (*subhāṣita*). It should be known to be the word of the Buddha if it is meaningful and not meaningless; if it is principled and not unprincipled; if it brings about the extinction of the afflictions of desire, hatred, and ignorance, and not their increase; and if it sets forth the qualities and benefits of nirvāṇa and not the qualities and benefits of saṃsāra.

The Buddha, out of his boundless compassion and omniscient wisdom, is said to teach what is most appropriate for each person. It is this infinite adaptability that provides the rationale for the proliferation of doctrine in the Buddha's name while continuing to claim a single and unchanging truth (variously identified across time and tradition). This apparent contradiction—and any contradiction is only apparent, since the Buddha must always be without contradiction—between multiple messages and a single truth is powerfully encapsulated in a common description of the Buddha's pedagogical powers. It is said that when the Buddha preached the dharma, each person in the audience heard a different discourse, perfectly suited to his or her intellect and interests, in his or her own native tongue, and felt that the Buddha was speaking to them alone. In fact, it is said, despite the length of the sermon that each person heard, the Buddha only uttered the syllable *a*, the first letter of the Sanskrit alphabet, and hence the source from which all language emanates. Each person heard what they wanted, or at least needed, to hear.

Buddhism thus appears as a tradition that is endlessly adaptable, with each added text, even those that have appeared in the modern period, seeming to have emerged somehow from the Buddha's monosyllabic pronouncement. Such a view would see each text as a contingent and mediated form of the Buddha's inexpressible enlightenment, adapted for the appropriate time and place. By these criteria, the question would thus seem to be: Is science *subhāṣita*, well-spoken? And if it is, what is rendered provisional thereby? Or put another way: is there anything that is not contained in the Buddha's capacious consciousness, a consciousness described as at once all-knowing and unknowable?

• • •

Over the history of the Buddhism and Science discourse, Buddhism has been identified in a variety of ways. Yet it has generally been the case that, regardless of the differences among the various Buddhisms that have been paired with various Sciences, they share a rather spare rationality, with the vast *imaginaire* of Buddhism largely absent; each is a Buddhism extracted from the Buddhist universe, a universe dense with deities.

In this book, I have attempted to describe some of the things that have been Buddhism over the centuries, things that do not fit so easily into the pairing with Science. As someone who has long sought to understand what Buddhism has been, it may seem that I am claiming the authority to pronounce what Buddhism should be, and what it cannot be. That has not been my purpose here. It may seem that I am afflicted with a certain nostalgia, trying somehow to keep Buddhism from changing, attempting to remythologize the demythologized. But this is impossible. My aim instead has been to document some of the ways that Buddhism has been represented as compatible with Science over the past 150 years. And my assumption is that this history will give pause to anyone who might have thought that Buddhism is modern, au courant, up-to-date with the latest scientific discoveries. In fact it has been, for far more than a century. It is my claim that to see Buddhism as ever modern comes at a cost, a price that many may consider well worth paying. But before paying that price, it is perhaps useful to recall those elements of Buddhism that are so starkly premodern, and to ask what is at stake in their loss. I chose to begin the previous chapter with a description of an important tantric meditation, in part because it contains elements that are so common throughout Buddhist thought and practice: invocation, incantation, visualization of the fantastic, and the sheer tedium of the text. By doing so, I seek to suggest to those who would reduce Buddhism to a philosophy or a science that these apparently exotic elements are essential to what Buddhism has been, and is. This is not to suggest that Buddhism has an essence, as so many apologists for Buddhism and Science have sought to identify. At the same time, those who would account for the adaptability of Buddhism by recourse to some facile claim that Buddhism has always been antiessentialist run the risk of allowing Buddhism to be everything, and nothing. It is neither. Perhaps the questions I have asked are meant merely to forestall further diminishing of the dimensions of the Buddha's aura, to keep the Buddha from becoming just a nice person.

But the Buddha should have the last word. In the Great Discourse on the Lion's Roar (*Mahāsīhanāda Sutta*), a famous text from the Pāli canon, the oldest collection of texts and, according to many, the collection that best represents his teachings, the Buddha declares, "Should anyone say of me: 'The recluse Gotama does not have any superhuman states, any distinction in knowledge and vision worthy of the noble ones. The recluse Gotama teaches a Dhamma [merely] hammered out by reasoning, following his own line of inquiry as it occurs to him'—unless he abandons that assertion and that state of mind and relinquishes that view, then as [surely as if he has been] carried off and put there, he will wind up in hell."[7]

NOTES

Preface

1 Ernest J. Eitel, *Buddhism: Its Historical, Theoretical and Popular Aspects in Three Lectures*, 2nd ed. (London: Trubner and Company, 1873), pp. 71–72. He continues:

> Again, the Buddhist idea of each world being subject to destruction by fire, in order to be re-constructed again in a similar form, cannot be repugnant to modern astronomers, who witnessed the disappearance of stars through blazing conflagrations and who believe in the existence of a resisting medium in space, which retarding every year the movement of every planet and every sun finally results in the dissolution of every globe, to give way—as Buddhism teaches us—to a new heaven and a new earth.
>
> Even some of the results of modern geology may be said to have been intuitively divined in Buddhism. For the Buddhists knew the interior of the earth to be in an incandescent state, they spoke of the formation of each earth as having occupied successive periods of incalculable duration, they strongly intimated that we are walking on catacombs of dead generations, that we subsist on a world resting on worlds vanished.

2 Eitel describes the formation of the Buddhist universe at ibid., pp. 42–44 and 47–49. The locus classicus for this description in Buddhist literature is the *Abhidharmakośa* (Treasury of Knowledge) by the fourth-century Indian scholar

Vasubandhu. For a translation of the relevant section, see Leo M. Pruden, trans., *The Abhidharmakośabhāṣyam by Louis de la Vallée Poussin* (Berkeley, CA: Asian Humanities Press, 1989), 2:451–74.

3 See Steven Kawaler and Joseph Veverka, "The Habitable Sun—One of Herschel's Stranger Ideas," *Journal of the Royal Astronomical Society of Canada* 75 (1981): 46–55.

Introduction

1 This is one of two statements about Buddhism attributed to Einstein that are widely quoted. The second is a variation of the first: "Buddhism has the characteristics of what would be expected in a cosmic religion for the future: It transcends a personal God, avoids dogmas and theology; it covers both the natural and the spiritual, and it is based on a religious sense aspiring from the experience of all things, natural and spiritual, as a meaningful unity." The attribution of these statements to Einstein is questionable. They do not appear, for example, in Thomas J. McFarlane, ed., *Einstein and Buddha: The Parallel Sayings* (Berkeley, CA: Seastone, 2002). When either of these of two statements is cited, either no source is provided or the source is identified as Helen Dukas and Banesh Hoffman, eds., *Albert Einstein: The Human Side* (Princeton, NJ: Princeton University Press, 1954), with no page number provided. In fact, the quotation does not appear in this volume, where the only mention of Buddhism or the Buddha is, "What humanity owes to personalities like Buddha, Moses, and Jesus ranks for me higher than all the achievements of the enquiring and constructive mind." See page 70; the German original appears on page 144.

2 For an insightful and judicious typology of the Buddhism and Science discourse, focusing especially on the past two decades, see José Ignacio Cabezón, "Buddhism and Science: On the Nature of the Dialogue," in *Buddhism and Science: Breaking New Ground*, ed. B. Alan Wallace (New York: Columbia University Press, 2003), pp. 35–68. For a useful study of recent scholarship on Science and Religion, see Peter Harrison, "'Science' and 'Religion': Constructing the Boundaries," *Journal of Religion* 86 (2006): 81–106. See also Richard K. Payne, "Buddhism and the Sciences: Historical Background, Contemporary Developments," in *Bridging Science and Religion*, ed. Ted Peters and Gaymon Bennett (Minneapolis, MN: Fortress Press, 2003), pp. 153–72.

3 Given my deficiencies in languages, I have relied largely on English and French sources, while providing my own translations from Tibetan for a number of the Buddhist materials. There are also some sources drawn from Chinese and Japanese, especially in chapter 1. It is important to note, however, that much of the literature on Buddhism and Science, regardless of the nationality or native language of its authors, was originally composed in English, providing some sense of the audience for whom it was intended.

4 See *The Encyclopædia Britannica with New American Supplement*, vol. 4 (New York: Werner Company, 1900), s.v. "Buddhism."

5 Eugène Burnouf, *Introduction à l'histoire du Buddhisme indien* (Paris: Imprimerie Royale, 1844), p. 337. Translation is from Eugène Burnouf, *Introduction to the History of Indian Buddhism*, trans. Ḱatia Buffetrille and Donald S.Lopez Jr. (Chicago: University of Chicago Press, forthcoming). Copyright © 2009 by The University of Chicago. All rights reserved.

6 Thomas H. Huxley, *Evolution and Ethics and Other Essays* (London: Macmillan and Company, 1894), pp. 68–69.

7 Indeed, some saw a closer link between the Indian Buddha and Europeans than between the Indian Buddha and East Asian Buddhists. In a 1913 work entitled *Buddhism & Science*, Paul Dahlke writes,

> The tongue in which the Buddha preached, taught, and thought, whether it was the Pāli itself or some dialect related to it, belongs to the Indo-Germanic stem. The root-words, the grammatical constructions, are akin to those found in European languages. Without any more said, we see how deep is the tie that binds us to the Buddha. . . .
>
> On the other hand, I should refer the intellectual derailment which the Buddha-thought has undergone in Tibet, China, and Japan, in no small measure to the lack of congruity that exists between the Indo-German and the Mongolian languages.

See Paul Dahlke, *Buddhism & Science* (London: Macmillan and Co., 1913), pp. 29 and 30.

8 On Blavatsky's turn to Asia, and especially to Brahmanism, see Mark Bevir, "The West Turns Eastward: Madame Blavatsky and the Transformation of the Occult Tradition," *Journal of the American Academy of Religion* 62, no. 3 (Fall 1994): 747–67.

9 Henry Steel Olcott, *Old Diary Leaves*, 2nd ser. (1878–93), 2nd ed. (Adyar, India: Theosophical Publishing House, 1928), pp. 167–68.

10 Ibid., pp. 168–69.

11 Henry S. Olcott, *The Buddhist Cathecism*, 44th ed. (Adyar, India: Theosophical Publishing House, 1947), p. 84.

12 Ibid., p. 88.

13 Ibid., p. 91.

14 Sir Edwin Arnold, *Seas and Lands* (New York: Longsman, Green, and Co., 1891), pp. 256–57.

15 Dharmapāla likely initially met Shaku Sōen in Sri Lanka, where the Japanese priest studied Theravāda Buddhism and Pāli from 1887 to 1889. For a useful introduction to the life and thought of Carus, see Martin Verhoeven's introduction to the new edition of Paul Carus, *The Gospel of Buddha: Compiled by Ancient Records* (Chicago: Open Court Publishing Company, 2004). On some of Carus's American contemporaries in the conversation on Buddhism and Science, see Thomas A. Tweed, *The American Encounter with Buddhism, 1844–1914: Victorian Culture and the Limits of Dissent* (Bloomington: Indiana University Press, 1992), pp. 103–10.

16 Anagārika Dharmapāla, "Message of the Buddha," in *Return to Righteousness: A Collection of Speeches, Essays and Letters of Anagārika Dharmapāla*, ed. Ananda Guruge (Ceylon: Government Press, 1965), p. 27.

17 Ibid., p. 439.

18 Ibid., p. 443.

19 This description of Chinese Buddhism is drawn from the standard work on the subject, Holmes Welch, *The Buddhist Revival in China* (Cambridge, MA: Harvard University Press, 1968).

20 On the life and work of Taixu, see Don Alvin Pittman, *Toward a Modern Chinese Buddhism: Taixu's Reforms* (Honolulu: University of Hawai'i Press, 2001).

21 His Eminence Tai Hsu, *Lectures in Buddhism* (Paris, 1928), pp. 11, 14. Others had a somewhat less grand view of his activities. In October 1928, Taixu was invited by the great French Sanskritist Sylvain Lévi to lecture at the Musée Guimet. According to one report, he delivered "a rambling, incoherent, and amateurish talk about the similarities between Buddhism, science, and Marxism" to a packed house, translated by a Chinese student who knew little about Buddhism. After the lecture, the French Sinologist (and friend of Debussy's) Louis Laloy remarked, "Nous nous sommes trompés." The *Bulletin de l'Association française des amis de l'orient* observed:

> We are grateful to the Chinese monk who has started the movement: at the same time we have noticed with some concern that there is perhaps a certain misunderstanding between him and his French listeners. . . . He is bent on showing [Europeans] that he is familiar with all the philosophies and sciences of the West . . . that there is no incompatibility between Buddhism and science. It is at least forty years since the conflict between science and religion ceased to interest anyone in Europe, but it is possible that his arguments will prove striking to Asians that have been Westernized, that is, Americanized. In short, it would seem that His Eminence T'ai-hsü has not taken into account the mentality of the countries to which he wants to bring the light of the law.

See Welch, *The Buddhist Revival in China*, pp. 59–60; see also pp. 65–66.

22 His Eminence Tai Hsu, *Lectures in Buddhism*, pp. 43, 47–48. A somewhat similar view, apparently uninfluenced by Taixu, was expressed sixty years later by the prominent Sinhalese monk Walpola Rahula in an address at the University of Kelaniya in Sri Lanka. He cautioned against seeking to make religions modern and accessible by validating religious truths with science, and notes that "we have almost become creatures or slaves of science and technology. Soon we shall be worshipping it. Early symptoms of it are that we tend to seek support from science to prove the validity of our religions, to justify them and to make them up-to date, respectable and acceptable. It is not surprising, therefore, that some well-intentioned Buddhist monks as well as some Buddhist laymen are making an ill advised effort to prove that Buddhism is a scientific religion." Rahula does not dispute for a moment that the Buddha clearly and precisely taught the latest scientific worldview, but holds that "to put on the same footing, the Buddha and a philosopher or a scientist, however celebrated he may be, is a gross disrespect to the Great Teacher." See Ven. Dr. Walpola Rahula, "Religion and Science," *Dharma Vijaya: Triannual Publication of Dharma Vijaya Buddhist Vihara Los Angeles* 2, no. 1 (May 1989): 10, 14. I am grateful to Martin J. Verhoeven for providing me with a copy of this article. See also Martin J. Verhoeven, "Buddhism and Science: Probing the Boundaries of Faith

and Reason," *Religion East & West: Journal of the Institute for World Religions* 1 (June 2001): 77–97.

23 Paul Dahlke writes, "The Buddha-teaching is a pure intuition, is *the* intuition, and proves itself such in that any attempt to treat of it after the methods of science, to master it inductively, is impossible." See Dahlke, *Buddhism & Science*, p. 84.

24 The "spirit" of the New Buddhism, as well as its close ties to the expansionist policies of the Meiji government, is expressed in the concluding paragraphs of an address entitled "History of Buddhism and Its Sects in Japan," delivered by the Shingon priest Toki Hōryū (1854–1923) on September 14, 1893, to the World's Parliament of Religions in Chicago.

> The present Japanese Buddhism has passed several hundred years since the last change. The past experience points out to us that it is time to remodel the Japanese Buddhism—that is, the happy herald is at our gates informing us that the Buddhism of the perfected intellect and emotion, synthesizing the ancient and modern sects, is now coming.
>
> The Japanese Buddhists have many aspirations, and at the same time great happiness, and we can not but feel rejoiced when we think of the probable result of this new change by which the Buddhism of great Japan will rise and spread its wings under all heaven as the grand Buddhism of the whole world.

See Horin Toki [Toki Hōryū], "History of Buddhism and Its Sects in Japan," in *Neely's History of the Parliament of Religions and Religious Congresses at the World's Columbian Exposition*, ed. Walter R. Houghton, 4th ed. (Chicago: F. Tennyson Neely, 1894), p. 226.

On Meiji policies regarding Buddhism and Buddhist responses, see James Edward Ketelaar, *Of Heretics and Martyrs in Meiji Japan* (Princeton, NJ: Princeton University Press, 1990), Richard Jaffe, *Neither Monk nor Layman: Clerical Marriage in Modern Japanese Buddhism* (Princeton, NJ: Princeton University Press, 2001), and Brian Victoria, *Zen at War* (New York: Weatherhill, 1997).

25 Shaku Soyen, "The Law of Cause and Effect, as Taught by the Buddha," in *Neely's History of the Parliament of Religions and Religious Congresses at the World's Columbian Exposition*, ed. Walter R. Houghton, 4th ed. (Chicago: F. Tennyson Neely, 1894), p. 378. Shaku Sōen was not the only member of the Japanese delegation to the World's Parliament to invoke science in his address. The Shingon priest Toki Hōryū stated, "In the Nirvana-Sutra it is said, 'All things have the nature of Buddha.' If the nature of all things is explained by mental science, biology, etc., it will be ascertained that the idea taught in the Nirvana-Sutra of the uniform nature in all things is true." See Horin Toki, "Buddhism in Japan," in *The World's Parliament of Religions: An Illustrated and Popular Story of the World's First Parliament of Religions, Held in Chicago in Conjunction with the Columbian Exposition of 1893*, ed. John Henry Barrows (Chicago: Parliament Publishing Company, 1893), 1:546.

26 A little more than a decade later, W. S. Lilly wrote in his essay "The Message of Buddhism to the Western World," "The truth is there is really one sole dogma of

Buddhism—that the whole universe is under one and the self-same law of causation which is ethical. That, it regards as the root of the matter; and so long as men hold fast to this prime verity, it views with indulgence the superfluous beliefs—*Aberglaube*, as the Germans say—in which they are led to indulge, by circumstances of time and place." See W. S. Lilly, "The Message of Buddhism to the Western World," *The Fortnightly Review*, n.s., 78 (July–December 1905): 209.

27 Huxley, *Evolution and Ethics and Other Essays*, p. 60.

28 Quoted in Shokin Furuta, "Shaku Sōen: The Footsteps of a Modern Japanese Zen Master," in "The Modernization of Japan," special issue, *Philosophical Studies of Japan* 7 (Tokyo: Japan Society for the Promotion of Science, 1967), p. 76. The biography of Sōen presented here is drawn largely from this source.

29 See Houghton, ed., *Neely's History of the Parliament of Religions and Religious Congresses at the World's Columbian Exposition*, pp. 797–98. For a fascinating study of the Japanese delegation, see Judith Snodgrass, *Presenting Japanese Buddhism to the West: Orientalism, Occidentalism, and the Columbian Exposition* (Chapel Hill: University of North Carolina Press, 2003).

30 Daisetz Teitaro Suzuki, *Outlines of Mahayana Buddhism* (Chicago: Open Court Publishing Company, 1908), p. 40.

31 Ibid., pp. 97–98. Half a century later, Suzuki would come to modify his views on religion and science. Writing about Paul Carus in 1959, he recalled, "What impressed Dr. Carus and Mr. Hegeler about Buddhism was the fact that Buddhism was singularly free from such mythological elements." He adds, "I now think that a religion based solely on science is not enough. There are certain 'mythological' elements in every one of us, which cannot be altogether lost in favor of science. This is a conviction I have come to. By this I do not mean that a religion is not to undergo a rigorous purging of all its 'impure' elements." See Daisetz Teitaro Suzuki, "A Glimpse of Paul Carus," in *Modern Trends in World Religions: Paul Carus Memorial Symposium*, ed. Joseph Kitagawa (LaSalle, IL: Open Court Publishing Company, 1959), p. x.

32 A. B. Keith, *Buddhist Philosophy in India and Ceylon* (Oxford: Clarendon Press, 1923), pp. 3–4.

33 See, for example, Daisetz Teitaro Suzuki, Erich Fromm, and Richard De Martino, *Zen Buddhism and Psychoanalysis* (New York: Harper, 1960), and Alan Watts, *Psychotherapy, East and West* (New York: Pantheon Books, 1961). Theravāda claims also continued during this period. The Burmese Buddhist monk U Thittila, playing on the multivalence of the word *dharma* (Pāli: *dhamma*), declared:

> All the teachings of the Buddha can be summed up in one word: Dhamma. . . . Dhamma, this law of righteousness, exists not only in a man's heart and mind, it exists in the universe also. All the universe is an embodiment and revelation of Dhamma. When the moon rises and sets, the rains come, the crops grow, the seasons change, it is because of Dhamma, for Dhamma is the law residing in the universe which makes matter act in the ways revealed by the studies of modern science in physics, chemistry, zoology, botany, and astronomy. Dhamma is the true nature of every existing thing, animate and inanimate.

U Thittila, "The Fundamental Principles of Theravada Buddhism," in *The Path of the Buddha: Buddhism Interpreted by Buddhists*, ed. Kenneth W. Morgan (New York: Ronald Press Company, 1957), p. 67.

Also relevant in the postwar period, especially in Japan, was the question, raised by Buddhists and others, of the use and misuse of science in the wake of the bombings of Hiroshima and Nagasaki. The Buddhist writing on this topic, much of which has yet to be translated into English, represents an important element of the Buddhism and Science discourse in the 1950s and 1960s. The writings of D. T. Suzuki on science, composed both in Japanese and English over the course of more than half a century, merit their own study. The complicity of Japanese Buddhists (including Suzuki) in the Pacific War has been a source of both repentance and recrimination. See, for example, Brian Daizen Victoria, *Zen at War*, 2nd ed. (Lanham, MD: Rowman and Littlefield, 2006), James W. Heisig and John C. Maraldo, eds., *Rude Awakenings: Zen, the Kyoto School, and the Question of Nationalism* (Honolulu: University of Hawai'i Press, 1995), and Jamie Hubbard and Paul L. Swanson, eds., *Pruning the Bodhi Tree: The Storm over Critical Buddhism* (Honolulu: University of Hawai'i Press, 1997).

34 Fritjof Capra, *The Tao of Physics: An Exploration of the Parallels between Modern Physics and Eastern Mysticism*, 3rd ed., updated (Boston: Shambhala, 1991), p. 19. The reference to Don Juan is probably significant, since Capra notes in the preface, "In the beginning, I was helped on my way by 'power plants' which showed me how the mind could flow freely; how spiritual insights come on their own, without any effort, emerging from the depth of consciousness." See p. 12.

35 Ibid., p. 36.

36 Ibid., p. 130.

37 Ibid., p. 93.

38 Ibid., p. 97.

39 See Daisetsu Teitaro Suzuki, *Aśvaghoṣa's Discourse on the Awakening of Faith in the Mahayana* (Chicago: Open Court Publishing Co., 1900), and Timothy Richard, *The Awakening of Faith in the Mahāyāna Doctrine—the New Buddhism* (Shanghai: Christian Literature Society, 1907). A study of the precise role of this text in the discourse of Buddhism and Science is a desideratum.

40 For a study of Suzuki's place in the history of Zen in the West, see Robert Sharf, "D. T. Suzuki and the Zen of Japanese Nationalism," in *Curators of the Buddha: The Study of Buddhism under Colonialism*, ed. Donald S. Lopez Jr. (Chicago: University of Chicago Press, 1995), pp. 107–60. For his presentation of Zen, Capra draws also upon the works of Alan Watts and the American Zen teacher Philip Kapleau, each of whom had a complicated relationship to the Zen traditions of Japan. On Kapleau, see Robert H. Sharf, "Sanbokyodan: Zen and the Way of the New Religions," *Japanese Journal of Religious Studies* 22, nos. 3–4 (1995): 417–58.

41 Capra, *The Tao of Physics*, p. 303.

42 Ibid., p. 340.

43 His Holiness the Dalai Lama, *Opening the Eye of New Awareness*, rev. ed., trans. and with an introduction by Donald S. Lopez Jr. (Boston: Wisdom Publications, 1999), p. 21.

44 Buddhism has not always been positively portrayed in this regard, as reported in *The Dublin Review* of 1875.

> Buddhism agrees with the moderns in considering this visible universe as the only reality, and as neither beginning nor ending ever; it is further consistent with Hegel in alternately declaring that nothing is, and that the sensible is somehow objective. But note how widely different are the conclusions drawn from identical premises. The Eastern withdraws himself into contemplation, subdues every moment of passion, and longs vehemently for one only consummation, that the semblance of life, which he has, may be taken from him, and pain may cease with annihilation. The Western cannot learn this hopeless asceticism; he believes that the cycle of things is an evolution from mere potentiality to some large perfection; he worships progress, and looks out for the means of advancing society and mankind of its onward path. He does not esteem life to be evil, though he must admit pain and disappointment to be the common lot, and therefore his lot also.

"Modern Society and the Sacred Heart," *Dublin Review*, n.s., 25 (July–October 1875): 17–18.

45 See *The Encyclopædia Britannica with New American Supplement,* vol. 4, s.v. "Lāmāism." On Victorian comparisons between Tibetan Buddhism and Roman Catholicism, see Donald S. Lopez Jr., *Prisoners of Shangri-La: Tibetan Buddhism and the West* (Chicago: University of Chicago Press, 1998), especially pp. 15–45.

46 His Holiness the Dalai Lama, *Opening the Eye of New Awareness*, p. 107.

47 A small and relatively random selection of relevant works in this regard (none of which deals with Buddhism) might include Steven Shapin, *The Scientific Revolution* (Chicago: University of Chicago Press, 1996); Peter J. Bowler and Iwan Rhys Morus, *Making Modern Science: A Historical Survey* (Chicago: University of Chicago Press, 2005); Gillian Beer, *Open Fields: Science in Cultural Encounter* (Oxford: Oxford University Press, 1996); Bernard Lightman, ed., *Victorian Science in Context* (Chicago: University of Chicago Press, 1997); Alan G. Gross, *The Rhetoric of Science* (Cambridge, MA: Harvard University Press, 1996); Jennifer Michael Hecht, *The End of the Soul: Scientific Modernity, Atheism, and Anthropology in France* (New York: Columbia University Press, 2003); Peter J. Bowler, *Reconciling Science and Religion: The Debate in Early-Twentieth-Century Britain* (Chicago: University of Chicago Press, 2001); Gyan Prakash, *Another Reason: Science and the Imagination of Modern India* (Princeton, NJ: Princeton University Press, 1999); Olav Hammer, *Claiming Knowledge: Strategies of Epistemology from Theosophy to the New Age* (Leiden: E. J. Brill, 2004); and John Brooke and Geoffrey Cantor, *Reconstructing Nature: The Engagement of Science and Religion* (Oxford: Oxford University Press, 2000), which contains a chapter on Capra.

48 Quoted in Dian J. Land, "Scientists Meet with Dalai Lama to Study Meditation," *Research News & Opportunities in Science and Theology* 1, no. 11/12 (July–August 2001): 2.

49 Lilly, "The Message of Buddhism to the Western World," 208–9.

Chapter 1

The title of this chapter alludes to the 1967 song by Donovan Leitch, "There Is a Mountain," with the lyrics, "First there is a mountain, then there is no mountain, then there is." This in turn is an allusion to the famous saying of the Tang dynasty Chan master Qingyuan weixin.

> Before I had studied Chan for thirty years, I saw mountains as mountains, and rivers as rivers. When I arrived at a more intimate knowledge, I came to the point where I saw that mountains are not mountains, and rivers are not rivers. But now that I have got its very substance I am at rest. For it's just that I see mountains once again as mountains, and rivers once again as rivers.

This passage was immortalized in two classic works for the European and American fascination with Zen. It presumably first appeared in English in D. T. Suzuki's *Essays in Zen Buddhism (First Series)*, first published by Rider for the Buddhist Society of London in 1926. Suzuki's translation reads, "Before a man studies Zen, to him mountains are mountains and waters are waters; after he gets an insight into the truth of Zen through the instruction of a good master, mountains to him are not mountains and waters are not waters; but after this when he really attains to the abode of rest, mountains are once more mountains and waters are waters." See Daisetz Teitaro Suzuki, *Essays in Zen Buddhism (First Series)* (New York: Grove Press, 1949), p. 24. The passage also appears in Alan Watts's 1957 *The Way of Zen* (from which the translation above is adapted). See Alan W. Watts, *The Way of Zen* (New York: Pantheon Books, 1951), p. 126. The passage entered the discourse of Buddhism and Science when it was cited by Fritjof Capra in his *The Tao of Physics*, first published in 1975. See Fritjof Capra, *The Tao of Physics: An Exploration of the Parallels between Modern Physics and Eastern Mysticism*, 3rd ed., updated (Boston: Shambhala, 1991), p. 124.

The melody of the Donovan song became the basis of "Mountain Jam" by the Allman Brothers Band, performed most famously during their concerts at the Fillmore East in New York in March 1971 and released in 1972 on the album *Eat a Peach*.

1 *A Full Account of the Buddhist Controversy, held at Pantura, in August, 1873. By the "Ceylon Times" Special Reporter: with the Addresses Revised and Amplified by the Speakers* (Colombo, Sri Lanka: Ceylon Times Office, 1873), pp. 1 and 2. For studies of relations between Buddhist monks and Christian missionaries in Sri Lanka during the nineteenth century, see R. F. Young and G. P. V. Somaratna, *Vain Debates: The Buddhist-Christian Controversies of Nineteenth-Century Ceylon* (Vienna: Publications of the De Nobili Research Library, 1996), and Elizabeth J. Harris, *Theravāda Buddhism and the British Encounter: Religious, Missionary, and Colonial Experience in Nineteenth-Century Sri Lanka* (London: Routledge, 2006). On the eighteenth century, see Anne M. Blackburn, *Buddhist Learning and Textual Practice in Eighteenth-Century Lankan Monastic Culture* (Princeton, NJ: Princeton University Press, 2001).

2 *A Full Account of the Buddhist Controversy, held at Pantura,* pp. 10–11.

3 Ibid., p. 13.

4 Ibid., pp. 2–3.

5 Ibid., p. 20.

6 Ibid., p. 42.

7 The story is recounted in Bkras mthong thub bstan chos dar, *Dge 'dun chos 'phel gyi lo rgyus* (Dharamsala, India: Library of Tibetan Works and Archives, 1980), p. 59.

8 *A Full Account of the Buddhist Controversy, held at Pantura,* p. 60.

9 Although typically translated as "aeon," the Sanskrit term *kalpa* is a length of time that cannot be measured by any standard unit. According to one description, one should imagine a great stone cube, a *yojana* (or some 5 miles) in height, breadth, and depth. Imagine, again, that once every century, someone were to wipe that great stone cube with a piece of soft Banaras silk. The amount of time it would take to thus wear away that stone cube is one *kalpa.*

10 For a translation of the relevant section of the *Abhidharmakośa*, see Leo M. Pruden, trans., *The Abhidharmakośabhāṣyam by Louis de la Vallée Poussin* (Berkeley, CA: Asian Humanities Press, 1989), 2:451–74. For other descriptions of the Buddhist universe, see Akira Sadakata, *Buddhist Cosmology: Philosophy and Origins* (Tokyo: Kōsei Publishing, 1997), Lati Rinpoche et al., *Meditative States in Tibetan Buddhism* (London: Wisdom Publications, 1983), pp. 23–47, and Jamgön Kongtrul Lodrö Tayé, *Myriad Worlds: Buddhist Cosmology in Abhidharma, Kālacakra and Dzog-chen* (Ithaca, NY: Snow Lion Publications, 1995), pp. 107–45.

11 Francis Xavier, *The Letters and Instructions of Francis Xavier*, trans. M. Joseph Costelloe, SJ (St. Louis: Institute of Jesuit Sources, 1992), p. 334. See also Masahiko Okada, "Vision and Reality: Buddhist Cosmographic Discourse in Nineteenth Century Japan" (Ph.D. diss., Stanford University, 1997) p. 112.

12 Tominaga Nakamoto, *Emerging from Meditation*, trans. with an introduction by Michael Pye (London: Gerald Duckworth & Co., Ltd, 1990), pp. 88–89, 93. For a useful discussion of Tominaga, see James Edward Ketelaar, *Of Heretics and Martyrs in Meiji Japan: Buddhism and Its Persecution* (Princeton, NJ: Princeton University Press, 1990). For another translation of the last portion of the passage cited here, see p. 24.

13 The description of Fumon Entsū and his contemporaries is drawn from Okada, "Vision and Reality," p. 71.

14 Ibid., pp. 75–76.

15 Ibid., p. 123.

16 Ibid., pp. 168–69.

17 Ibid., p. 167.

18 Ibid., pp. 187, 190.

19 Ibid., p. 197.

20 Guy Tachard, *A Relation of the Voyage to Siam: Performed by Six Jesuits, Sent by the French King, to the Indies and China, in the year, 1685: With their Astrological Observations, and their Remarks of Natural Philosophy, Geography, Hydrography, and History Published in the Original, by the Express Orders of His Most Christian Majesty; and Now made English, and Illustrated with Sculptures* (London, 1688), pp. 284–86.

21 Ibid., p. 289.

22 See Young and Somaratna, *Vain Debates*, pp. 57–58.

23 Hardy's motivation for his scholarship on Buddhism is articulated in the preface to his 1853 *A Manual of Budhism in Its Modern Development*.

> By the messengers of the cross, who may succeed me in the field in which it was once my privilege to labour, this Manual will be received, I doubt not, as a boon; as it will enable them more readily to understand the system they are endeavouring to supersede, by the establishment of the Truth. I see before me, looming in the distance, a glorious vision, in which the lands of the east are presented in majesty: happy, holy, and free. I may not; I dare not attempt to describe it; but it is the joy of my existence to have been an instrument, in a degree however feeble, to bring about this grand consummation.

See Robert Spence Hardy, *A Manual of Budhism in Its Modern Development* (London: Partridge and Oakey, 1853), p. xiii.

24 Robert Spence Hardy, *The Sacred Books of the Buddhists Compared with History and Science* (Colombo, Sri Lanka: Wesleyan Mission Press, 1863), pp. 71–73. The work was published in London in an expanded edition and modified title in 1866. See Robert Spence Hardy, *The Legends and Theories of the Buddhists Compared with History and Science With Introductory Notices of the Life and System of Gotama Buddha.*

In *Buddhism and Its Christian Critics*, first published in 1894, Paul Carus writes of Hardy, "Mr. Hardy speaks of 'the errors of Buddhism that are contrary to fact as taught by established and uncontroverted science' (p. 135), but appears to reject science whenever it comes into collision with a literal interpretation of Christian doctrines. Buddhism is to him a fraud, Christianity divine revelation." See Paul Carus, *Buddhism and Its Christian Critics* (Chicago: Open Court Publishing Company, 1897), p. 265.

25 Hardy, *The Sacred Books of the Buddhists Compared with History and Science*, pp. 69–70.

26 Ibid., p. 93.

27 Ibid., p. 101.

28 Ibid., pp. 77–78.

29 *A Full Account of the Buddhist Controversy, held at Pantura*, p. 60.

30 Ibid., pp. 66–67.

31 On Morrison, see the entry in the *Oxford Dictionary of National Biography* and Katherine Anderson, "The Weather Prophets: Science and Reputation in Victorian Meteorology," *History of Science* 37 (1999): 179–216.

32 *A Full Account of the Buddhist Controversy, held at Pantura*, pp. 67–68.

33 See *Zhen xianshi lun zong yi lun (zhong)*, in *Taixu dashi quanshu*, ed. Yinshun, vol. 19, CD-ROM (Xinzhu, Taiwan: Caituan faren Yinshun wenjiao jijinhui, 2005), pp. 228–33, 263. I am grateful to Scott Pacey for bringing these passages to my attention and providing a translation.

34 Dalai Lama, *The Universe in a Single Atom: The Convergence of Science and Spirituality* (New York: Morgan Books, 2005), p. 5.

35 Drang po Dharma, "'Jigs rten ril mo 'am zlum po," in *Yul phyogs so so'i gsar 'gyur me long*, June 28, 1938, p. 11.

36 On the question of Tibetans' views on European geography during this period, one finds an amusing passage in Heinrich Harrer's memoir, *Seven Years in Tibet*:

> A year later [1948] Professor Tucci, the famous Tibetologist, arrived from Rome. This was his seventh visit to Tibet, but his first to Lhasa. He was reckoned to be the greatest authority on the history and civilisation of Tibet, and had translated numerous Tibetan books as well as publishing a number of original works. He always astonished Chinese, Nepalese, Indians and Tibetans by his knowledge of the history of their countries. I often met him at parties, and once before a large gathering he put me in a very false position by taking sides with the Tibetans against me in an argument about the shape of the earth. In Tibet the traditional belief is that the earth is a flat disk. This was being argued at the party and I stood up for the spherical theory. My arguments seemed to be convincing the Tibetans and I appealed to Professor Tucci to support me. To my great surprise he took up a sceptical attitude, saying that in his opinion all scientists ought continually to be revising their theories, and that one day the Tibetan doctrine might just as well prove to be true! Everybody chuckled as it was known that I gave lessons in geography. Professer Tucci stayed eight days in Lhasa and then went on a visit to the most famous monastery in Samye; after which he left the country, taking with him scientific specimens and many valuable books from the Potala Palace.

See Heinrich Harrer, *Seven Years in Tibet* (London: Harper Perennial, 2005), pp. 195–96. I am grateful to Isrun Engelhardt for calling this passage to my attention.

37 Hor khang bsod nams dpal 'bar, ed., *Dge 'dun chos 'phel gyi gsung rtsom* (Lhasa, Tibet: Bod ljongs bod yig dpe rnying dpe skrun khang, 1990), vol. 2 (Gangs can rig mdzod 11), p. 402. Translation is from Gendun Chopel, *In the Forest of Faded Wisdom: 104 Poems by Gendun Chopel, a Bilingual Edition*, ed. and trans. Donald S. Lopez Jr. (Chicago: University of Chicago Press, forthcoming). Copyright © 2009 by The University of Chicago. All rights reserved.

38 See Donald S. Lopez Jr., ed., *Buddhist Hermeneutics* (Honolulu: University of Hawai'i Press, 1988).

39 The Dalai Lama, *The Way to Freedom* (San Francisco: HarperSanFrancisco, 1994), p. 73. One might note that the Japanese monk Fukuda Gyōkai wrote in his 1878 *Shumisen ryakusetsu* (*Summary of Shumisen Theory*) that "the Buddha did not teach the Shumisen worldview for land survey or astronomical calculation." See Okada, "Vision and Reality," p. 169.

40 N. A. Jayawickrama, trans., *The Story of Gotama Buddha (Jātaka-nidāna)* (Oxford: Pali Text Society, 2002), p. 99.

41 The information in this paragraph is drawn from Padmanabh S. Jaini, "On the *Sarvajñatva* (Omniscience) of Mahāvira and the Buddha," originally published in *Buddhist Studies in Honour of I. B. Horner*, ed. Lance Cousins et al. (Dordrecht: D.

Reidel Publishing, 1974), pp. 71–90. The essay was reprinted in Padmanabh S. Jaini, *Collected Papers in Buddhist Studies* (Delhi: Motilal Banarsidass, 2001), pp. 97–121. The quotation at the end of the paragraph appears on page 115 of the latter volume.

42 Tenzin Gyatso, *Opening the Eye of New Awareness*, rev. ed., trans. and with an introduction by Donald S. Lopez Jr. (Boston: Wisdom Publications, 1999), pp. 96–97.

43 Translated fron the Tibetan, *Pramāṇavarttika* 2.31–33. See Dharmakīrti, *Pramāṇavarttika* in *Sde dge Tibetan Tripiṭaka Bstan Ḥgyur Preserved at the Faculty of Letters, University of Tokyo* (*Tshad ma*) 1, 276 (Ce) (Tokyo: Sekai Seiten Kanko Kyokai, 1981), p. 54 (15/108b5–7). For the Sanskrit (where the verses are 1.33–35), see Ram Chandra Pandeya, ed., *The Pramāṇavarttikam of Ācārya Dharmakīrti* (Delhi: Motilal Banarsidass, 1989), p. 11.

44 Dalai Lama, *The Universe in a Single Atom*, p. 79.

45 Ibid., p. 80.

46 See E. Obermiller, trans., *History of Buddhism (Chos ḥbyung) by Bu-ston*, part 2, *The History of Buddhism in India and Tibet* (Heidelberg: Harrassowitz, 1932), pp. 140–41.

47 Ernest J. Eitel, *Buddhism: Its Historical, Theoretical and Popular Aspects in Three Lectures*, 2nd ed. (London: Trubner and Company, 1873), p. 45.

Chapter 2

1 Bundesarchiv Berlin, R 5101/23400, fols. 98–99, arrived October 27, 1937. The archived letter is in German; neither the Chinese original nor the name of the translator has been identified. The four virtues mentioned in this letter are "Mitleid (mit der Not des Nächsten), Einordnung (in die sociale Rangordnung), Wirken (für Besserung) und Mut (Widerstände zu brechen)." Without the original Chinese-language document, it is unclear to which Buddhist virtues this list is meant to refer. I am grateful to Isrun Engelhardt, for calling this letter to my attention and for providing the translation from German.

2 This story appears, among other places, in the first Brahmāsaṃyutta of the *Saṃyuttanikāya*. For an English translation, see Bhikkhu Bodhi, *The Connected Discourses of the Buddha: A New Translation of the Saṃyutta Nikāya* (Boston: Wisdom Publications, 2000), 1:231–33.

3 In this chapter, the term *caste* will be used to render the Sanskrit term *varṇa*, referring to the classical fourfold division of society, rather than to *jāti*, the term commonly translated as "caste" in the modern period.

4 See Patrick Olivelle, *Manu's Code of Law: A Critical Edition and Translation of the Mānava-Dharmaśāstra* (Oxford: Oxford University Press, 2005), p. 128.

5 The statement occurs in the *Aggañña Sutta* of the *Dīghanikāya*. For an English translation, see Maurice Walshe, *The Long Discourses of the Buddha: A Translation of the Dīgha Nikāya* (Boston: Wisdom Publications, 1995), p. 408.

6 Among the large number of works that set forth this view, one that might be cited as typical of the genre is a pamphlet written by two distinguished Pāli scholars for a UNESCO series on the attitudes of the world religions toward racism, called The Race Question and Modern Thought. See G. P. Malalasekara and K. N. Jayatilleke, *Buddhism and the Race Question* (Paris: UNESCO, 1958).

7 Thomas W. Rhys Davids, *Dialogues of the Buddha* (London: H. Frowde, 1899–1921), 2:197.

8 See Uma Chakravarti, *The Social Dimensions of Early Buddhism* (New Delhi: Munshiram Manoharlal, 1996), pp. 98–101.

9 The statement occurs in the *Ambaṭṭha Sutta* of the *Dīghanikāya*. For an English translation, see Walshe, *The Long Discourses of the Buddha*, pp. 117–18.

10 See, for example, N. A. Jayawickrama, trans., *The Story of Gotama Buddha* (*Jātaka-nidāna*) (Oxford: Pali Text Society, 2002), p. 65. The *Mahāpadāna Sutta* identifies the castes of previous buddhas as either brahman or kṣatriya.

11 *The Group of Discourses* (*Sutta-nipāta*), rev. trans. K. R. Norman (Oxford: Pali Text Society, 1992), 2:71.

12 See Chakravarti, *The Social Dimensions of Early Buddhism*, pp. 124–25.

13 The story occurs in the *Pācittiya*. See ibid., p. 184. An interesting attempt to inflate the size of the lower-caste membership in the saṅgha is offered by G. P. Malalasekara and K. N. Jayatilleke, who, noting that only 8 percent of the nuns mentioned in the *Songs of the Sisters* (*Therīgāthā*) are "base-born," conclude, "Perhaps it would be nearer the truth to say that if 8 ½ per cent of the contributed poems were composed by and express the religious joy that the members of the despised castes felt on joining the Order and realizing the fruits of the training that it gave, then the actual percentage of the women of 'low' birth in the Order would have been much larger, since the social class from which they were drawn was mostly illiterate." See Malalasekara and Jayatilleke, *Buddhism and the Race Question*, p. 56.

14 Jayawickrama, *The Story of Gotama Buddha*, p. 121.

15 The statement appears in the *Cankī Sutta* of the *Majjhimanikāya*. For an English translation, see Bhikkhu Ñāṇamoli and Bhikkhu Bodhi, trans., *The Middle Length Discourses of the Buddha: A New Translation of the Majjhima Nikāya* (Boston: Wisdom Publications, 1995), pp. 776–77.

16 For an English translation, see Walshe, *The Long Discourses of the Buddha*, p. 413.

17 See ibid., pp. 413–14.

18 See Chakravarti, *The Social Dimensions of Early Buddhism*, pp. 39–44.

19 The passage is found in the *Bhayabherava Sutta* of the *Majjhimanikāya*. The English translation is adapted from Ñāṇamoli and Bodhi, *The Middle Length Discourses of the Buddha*, p. 105. It is noteworthy that the translators choose to render the Pāli term *vaṇṇa* (*varṇa* in Sanskrit) as "appearance" rather than "caste." For an analysis of this passage, see Donald S. Lopez Jr., "Memories of the Buddha," in *In the Mirror of Memory: Mindfulness and Remembrance in Indic and Tibetan Buddhism*, ed. Janet Gyatso (Albany: SUNY Press, 1992), pp. 21–45.

20 For a study of the arguments put forth by Buddhist logicians, see Vincent Eltschinger, *"Caste" et philosophie Bouddhique: Continuité de quelques arguments Bouddhiques contre le traitment réaliste des dénominations sociales* (Vienna: Arbeitskreis für Tibetische und Buddhistische Studien Universität Wien, 2000). For a brief survey of early Buddhist attitudes toward caste, see Y. Krishnan, "Buddhism and the Caste System," *Journal of the International Association of Buddhist Studies* 9, no. 1 (1986): 71–83.

21 See Donald S. Lopez Jr., "Paths Terminable and Interminable," in *Path to Liberation: The Mārga and Its Transformations in Buddhist Thought*, ed. Robert E. Buswell Jr. and Robert M. Gimello (Honolulu: University of Hawai'i Press, 1992), pp. 147–58.

22 See Chakravarti, *The Social Dimensions of Early Buddhism*, p. 112.

23 See K. R. Norman, "Four Noble Truths," in *Collected Papers 2* (Oxford: Pali Text Society, 1991), pp. 210–23.

24 In commenting on *Abhidharmakośa* 6.2. For an English translation, see Leo M. Pruden, trans., *The Abhidharmakośabhāṣyam by Louis de la Vallée Poussin* (Berkeley, CA: Asian Humanities Press, 1989), 3:898.

25 For a useful overview of this politically charged controversy, see Thomas R. Trautmann, *The Aryan Debate* (New Delhi: Oxford University Press, 2005).

26 See G. P. Malalasekara, *Dictionary of Pāli Proper Names*, vol. 1 (New Delhi: Munshiram Manoharlal, 1998), s.v. "Ariya." The story appears at *Dhammapadaṭṭhakathā* 3.396–98.

27 See Madhav Deshpande, "What to Do with the *Anāryas*: Dharmic Discourses of Inclusion and Exclusion," in *Aryan and Non-Aryan in South Asia: Evidence, Interpretation and Ideology*, ed. Johannes Bronkhorst and Madhav M. Deshpande (Cambridge, MA: Harvard Oriental Series, 1999), pp. 117–24. Some linguists speculate that *mleccha*, with its odd *m-l* ligature, derives from the attempt of a Sanskrit speaker to represent what a non-Sanskritic language sounds like.

28 See ibid., pp. 118–19.

29 See the *Aṅgulimālasutta* of the *Majjhimanikāya*. For an English translation, see Ñāṇamoli and Bodhi, *The Middle Length Discourses of the Buddha*, pp. 710–17. Aṅgulimāla's statement has been repeated by monks to pregnant women over the centuries in the hope of assuring a successful birth.

30 Eugène Burnouf, *Introduction à l'histoire du Buddhisme indien* (Paris: Imprimerie Royale, 1844), p. 212. This and all subsequent translations from Burnouf are from Eugène Burnouf, *Introduction to the History of Indian Buddhism*, trans. Katia Buffetrille and Donald S. Lopez Jr. (Chicago: University of Chicago Press, forthcoming). Copyright © 2009 by The University of Chicago. All rights reserved.

31 Julius von Klaproth, *Asia Polyglotta* (Paris: A. Schubart, 1823), p. 122. See also Julian von Klaproth, "Vie de Bouddha d'après les livres mongols," *Journal Asiatique* 4 (1824): 10.

32 Edward Upham, *The History and Doctrine of Budhism, Popularly Illustrated With Notices of the Kappooism or Demon Worship and of the Bali or Planetary Incantations of Ceylon* (London: R. Ackermann, 1829), pp. 12–13.

It is important to note that perhaps the greatest authority on Indian religions of his day, Henry Thomas Colebrooke, remained silent on the issue. Colebrooke, who succeeded Sir William Jones on the bench of the Supreme Court in Calcutta and exceeded him in the knowledge of Sanskrit, published five influential and widely cited essays, "On the Philosophy of the Hindus," between 1823 and 1827. The last of these, read on February 3, 1827, was entitled "On Indian Sectaries" and dealt with the Jains and Buddhists. On caste, Colbrooke merely says, "The *Jainas* and *Bauddhas* I consider to have been originally Hindus; and the first-mentioned to be so still, because they recognized, as they yet do, the distinction of the four castes." See Henry

Thomas Colebrooke, "On the Philosophy of the Hindus. Part IV," *Transactions of the Royal Asiatic Society of Great Britain and Ireland*, vol. 1 (1827), p. 549.

33 Charles F. Neumann, "Buddhism and Shamanism," *Asiatic Journal and Monthly Register* 16 (1835): 124. For a later and more extended discussion of caste in Buddhism, see the discourse of the Wesleyan Methodist missionary to Ceylon Robert Spence Hardy in his 1853 *A Manual of Budhism*: R. Spence Hardy, *A Manual of Budhism, in its Modern Development* (reprint, Varanasi, India: Chowkhamba Sanskrit Series Office, 1967), pp. 65–85. There he describes caste as "a deadly incubus, exerting its power every moment throughout century after century, upon the minds of a great proportion of the people" (p. 83), and notes:

> That which gives to caste its real importance, and by which it is exhibited in its most repulsive aspect, is, however, held as firmly by the Budhists as the Brahmans; inasmuch as they teach that the present position of all men is the result of the merit or demerit of former births; a doctrine which, if true, would make the scorn with which the outcast is regarded a natural feeling, as he would be in reality a condemned criminal, undergoing the sentence that has been pronounced against him by a tribunal that cannot err in its decrees. (pp. 78–79)

34 William Jones, "On the Chronology of the Hindus," *Asiatick Researches* 2 (1801): 125.

35 Ibid., p. 121. A more modern translation of the passage is:

> Moved by deep compassion, you condemn the Vedic way
> That ordains animal slaughter in rites of sacrifice.
> You take from as the enlightened Buddha, Krishna.
> Triumph, Hari, Lord of the World!

See Barbara Stoler Miller, ed. and trans., *Love Song of the Dark Lord: Jayadeva's Gītagovinda* (New York: Columbia University Press, 1977), p. 71.

36 Jones, "On the Chronology of the Hindus," p. 123.

37 Ibid., p. 124.

38 Brian Houghton Hodgson had written an important, and largely overlooked, essay on the topic in 1829. See "A Disputation Respecting Caste by a Buddhist," *Transactions of the Royal Asiatic Society*, vol. 3 (1835): 160–69. In 1836 he wrote in the *Journal of the Asiatic Society of Bengal*, "Saugata [Buddhist] books treating on the subject of caste never call in question the antique fact of the fourfold division of the Hindu people, but only give a more liberal interpretation to it than the current Brahmanical one of the day." See Brian Houghton Hodgson, "Quotations from Original Sanskrit Authorities in Proof and Illustration of the Preceding Article," *Journal of the Asiatic Society of Bengal* 5 (1836): 31.

39 The important works that preceded the publication of Burnouf's *Introduction* were translations, especially those from Tibetan and Mongolian by Isaak Jakob Schmidt, from Pāli by James Turnour, and from Chinese by Jean-Pierre Abel-Rémusat. Burnouf cites each of these authors repeatedly. The annotations to

Abel-Rémusat's 1836 *Foĕ Kouĕ Ki*, his translation of Faxian's *Foguoji*, or *Record of Buddhist Kingdoms*, contain a great deal of information (and speculation) on the Buddha and his disciples.

40 Burnouf, *Introduction*, pp. 210, 211.

41 Ibid., p. 211.

42 Ibid., p. 212.

43 Ibid., p. 213.

44 Ibid., pp. 214–15.

45 Ibid., pp. 194–95.

46 Ibid., p. 213–14.

47 *The Life and Letters of the Right Honourable Friedrich Max Müller, Edited by His Wife*, 2 vols. (London: Longmans, Green, and Co., 1902), 1:34.

Although composed some decades after the period being considered, it is perhaps of some interest to briefly note the views of the great German biographer of the Buddha Hermann Oldenberg (1854–1920), who in his 1881 *Buddha: Sein leben, seine lehre, seine gemeinde*, considered "the historically untrue conception of Buddha as the victorious champion of the lower classes against a haughty aristocracy of birth and brain."

> We can quite understand how historical treatment in our times, which takes a delight in deepening its knowledge of religious movements by bringing into prominence or discovering their social bearings, have attributed to the Buddha the *rôle* of a social reformer, who is conceived to have broken the chains of caste and won for the poor and humble their place in the spiritual kingdom which he founded. But any one who attempts to describe Buddha's labours must, out of love for truth, resolutely combat the notion that the fame of such an exploit, in whatever way he may depict it to himself, belongs to Buddha. If any one speaks of a democratic element in Buddhism, he must bear in mind that the conception of any reformation of national life, every notion in any way based on the foundation of an ideal earthly kingdom, of a religious Utopia, was quite foreign to this fraternity. There was nothing resembling a social upheaval in India. Buddha's spirit was a stranger to that enthusiasm, without which no one can pose as the champion of the oppressed against the oppressor. Let the state and society remain what they are; the religious man, who as a monk has renounced the world, has no part in its care and occupations. Caste has no value for him, for everything earthly has ceased to affect his interests, but it never occurs to him to exercise his influence for its abolition or for the mitigation of the severity of its rules for those who have lagged behind in worldly surroundings.

See Hermann Oldenberg, *Buddha: His Life, His Order, His Doctrine*, trans. William Hoey (London: Williams and Norgate, 1882), pp. 153–54. As Philip Almond notes, perusing reviews of Oldenberg's work, in the 1880s there was no difficulty in portraying the Buddha as anticlerical, but it was wrong to conclude that he was a socialist. See Philip Almond, *The British Discovery of Buddhism* (Cambridge:

Cambridge University Press, 1988), pp. 75–76. A decade before Oldenberg, Henry Alabaster had observed, regarding the doctrine of karma, "The teaching on this point may be said to recognise the equality of all beings, at the same time that it provides against the mischievous results European Socialists draw from that doctrine; which it does, by declaring the incompatibility of intrinsic quality of being with actual difference of condition and advantages." See Henry Alabaster, *The Wheel of the Law* (London: Trubner and Company, 1871), p. xxxvi.

48 See Eltschinger, *"Caste" et philosophie Bouddhique*.

49 Burnouf, *Introduction*, pp. 205–6.

50 Léon Poliakov, *The Aryan Myth: A History of Racist and Nationalist Ideas in Europe* (New York: Basic Books, 1974), p. 238.

51 Cited in ibid., p. 233.

52 Joseph Arthur, le Comte de Gobineau, *Essai sur l'inégalité des races humaines*, 2nd ed. (Paris: Librairie de Firmin-Didot et Cie, 1884), 1:437. It is likely that the count drew this conclusion by reading a passage like the following in Burnouf:

> I have already emphasized, in the first section of this memorandum, the difference between the Buddhist teaching and that of the brahmans. This difference is entirely in the teaching, which had the effect of bringing truths, which previously were the share of the privileged castes, within the reach of all. It gives Buddhism a character of simplicity, and, from the literary point of view, of mediocrity, that distinguishes it in the most profound manner from Brahmanism. It explains how Śākyamuni was led to receive among his listeners persons whom the highest classes of society rejected. It accounts for his successes, that is to say, for the facility with which his doctrine spread and his disciples multiplied. Finally, it provides the secret of the fundamental modifications that the propagation of Buddhism must have brought to the Brahmanical constitution, and of the persecutions that the fear of change could not fail to attract to the Buddhists from the day they became strong enough to place in peril a political system founded principally on the existence and perpetuation of the castes. These facts are so intimately linked with one another that it is sufficient that once the first took place, the others, with time, developed in an almost necessary way. (Burnouf, *Introduction*, p. 194)

53 De Gobineau, *Essai sur l'inégalité des races humaines*, 1:440.

54 Ibid., 1:443.

55 Ibid., 1:442.

56 See Sir Edwin Arnold, *East and West: Being Papers Reprinted from the "Daily Telegraph" and Other Sources* (London: Longmans, Green, and Co., 1896), p. 311.

57 Cited in John Henry Barrows, ed., *The World's Parliament of Religions: An Illustrated and Popular Story of the World's First Parliament of Religions, Held in Chicago in Connection with the Columbian Exposition of 1893* (Chicago: Parliament Publishing Company, 1893), 1:95.

58 A biography of Dharmapāla remains a desideratum. A chronology of his life, quotations from newspaper articles about him, and excerpts from the file maintained on him by the British colonial administration of Ceylon are included in the introduction to Ananda Guruge, ed., *Return to Righteousness: A Collection of Speeches, Essays and Letters of Anagarika Dharmapala* (Colombo, Sri Lanka: Ministry of Education and Cultural Affairs, 1965). His life and work are discussed in many sources, several of which can be mentioned here. See Bhikshu Sangharakshita, *Anagarika Dharmapala: A Biographical Sketch*, The Wheel Publication no. 70–72 (Kandy, Sri Lanka: Buddhist Publication Society, 1964), and Balkrishna Govind Gokhale, "Anagarika Dharmapala: Toward Modernity through Tradition in Ceylon," in *Contributions to Asian Studies* 4 (1973): 30–39. For an Eriksonian analysis of his personality, see Gananath Obeyesekere, "Personal Identity and Cultural Crisis: The Case of Anagārika Dharmapāla of Sri Lanka," in *The Biographical Process: Studies in the History and Psychology of Religion*, ed. Frank E. Reynolds and Donald Capps (The Hague: Mouton, 1976), pp. 221–52. His legacy for the Sri Lankan saṅgha is discussed in H. L. Seneviratne, *The Work of Kings* (Chicago: University of Chicago Press, 1999). See also George D. Bond, *The Buddhist Revival in Sri Lanka* (Columbia: University of South Carolina Press, 1988), and Torkel Brekke, *Makers of Modern Indian Religion in the Late Nineteenth Century* (Oxford: Oxford University Press, 2002), pp. 86–115.

59 Max Müller, *The Languages of the Seat of War in the East with a Survey of the Three Families of Language, Semitic, Arian, and Turanian*, 2nd ed. (London: Williams and Norgate, 1855), p. 31. The "War in the East" of the title is the Crimean War. This work is also known simply as *A Survey of Languages*. The same statement appeared a year earlier in Müller's *Suggestions for the Assistance of Officers in Learning the Languages of the Seat of War in the East* (London: Longman, Bowman, Green, & Longmans, 1854), p. 32, with the robes being described as "classical" rather than "ceremonial." In his *Lectures on the Science of Language*, delivered at the Royal Institution in London in 1861, Müller had noted in passing that minute and careful comparison "enables us to class the idioms spoken in Iceland and Ceylon as cognate dialects." See F. Max Müller, *Lectures on the Science of Language Delivered at the Royal Institution of Great Britain in April, May, and June, 1861*, 2nd rev. ed. (New York: Charles Scribner, 1862), 1:63. In his citation of this passage, R. A. L. H. Gunawardana adds at the end of the passage "of the Aryan family of languages." This phrase does not appear in Müller's text, although it appears to be his implication.

60 James de Alwis, "On the Origin of the Sinhalese Language," *Journal of the Ceylon Branch of the Royal Asiatic Society* 4, no. 13 (1865–66): 143. The passage from Müller is cited on page 156.

61 Ibid., p. 151. De Alwis would provide a more detailed linguistic case for Sinhalese as an Aryan language in his "On the Origin of the Sinhalese Language," *Journal of the Ceylon Branch of the Royal Asiatic Society* 4, no. 14 (1867–70, part 1): 1–86. The use of the term *race* in this statement confirms George Stocking's observation that in the late nineteenth century, "'blood'—and by extension 'race'—included numerous elements that we could today call cultural; there was no clear line between cultural and physical elements or between social and biological heredity." See George W.

Stocking Jr., *Delimiting Anthropology: Occasional Essays and Reflections* (Madison: University of Wisconsin Press, 2001), p. 8.

62 Sir Henry Sumner Maine, *Village-Communities in the East and West: Six Lectures Delivered at Oxford to which are added other Lectures, Addresses and Essays* (New York: Henry Holt and Company, 1889), p. 209. Maine goes on to write:

> The assumption, it is true, that affinities between the tongues spoken by a number of communities are conclusive evidence of their common lineage, is one which no scholar would accept without considerable qualification; but this assumption has been widely made, and in quarters and among classes where the discoveries out of which it grew are very imperfectly understood. There seems to me no doubt that modern philology has suggested a grouping of peoples quite unlike anything that had been thought of before. (Ibid.)

63 See Vijaya Samaraweera, "Litigation, Sir Henry Maine's Writings and the Ceylon Village Communities Ordinance of 1871," in *Senarat Paranavitana Commemoration Volume*, ed. Leelananda Prematilleke, Karthigesu Indrapala, and J. E. van Lohuizende Leeuw, Studies in South Asian Culture 7 (Leiden: E. J. Brill, 1978), pp. 191–203.

64 Mahadeva Moreshwar Kunte, "Lecture on Ceylon" (Bombay, 1880), p. 9.

65 Lewis Henry Morgan, *Ancient Society; or, Researches in the Lines of Human Progress from Savagery, through Barbarism to Civilization* (London: Macmillan and Co., 1877), p. 40.

66 Guruge, *Return to Righteousness*, p. 479.

67 For a useful analysis of the Vijaya myth, see R. A. L. H. Gunawardana, "The People of the Lion: The Sinhala Identity in History and Historiography," in *Sri Lanka: History and the Roots of Conflict*, ed. Jonathan Spencer (London: Routledge, 1990), pp. 48–58.

68 For examinations of the role of Dharmapāla's rhetoric in the recent communal violence in Sri Lanka, see Kumari Jayawardena, "Some Aspects of Class and Ethnic Consciousness in Sri Lanka in the Late 19th and Early 20th Centuries," in *Ethnicity and Social Change in Sri Lanka: Papers Presented at a Seminar Organised by the Social Scientists Association December 1979* (Colombo: Social Scientists' Association, 1984), pp. 74–92. An edited version of this essay was published as "Class Formation and Communalism" in the journal *Race & Class* (in a special issue entitled "Sri Lanka: Racism and the Authoritarian State") 26, no. 1 (Summer 1984): 51–62. See also Peter Schalk, "Buddhistische Kampfgruppen in Sri Lanka," *Asien* 21 (October 1986): 30–62. On the negative response to the essay that appeared in *Ethnicity and Social Change in Sri Lanka*, see Serena Tennekoon, "Newspaper Nationalism: Sinhala Identity as Historical Discourse," in Spencer, ed., *Sri Lanka*, pp. 214–21.

69 See Guruge, *Return to Righteousness*, p. 541.

70 Ibid., p. 518.

71 Ibid., p. 232.

72 Ibid., p. 460.

73 Ibid., p. 459.

74 Ibid., p. 447.

75 Ibid., p. 24.

76 Ibid., p. 57.

77 Ibid., p. 443.

78 Müller, *The Languages of the Seat of War in the East*, p. 29. For an illuminating study of the context of Müller's declaration and its legacy, see Thomas R. Trautmann, *Aryans and British India* (Berkeley: University of California Press, 1997), pp. 172–216.

79 Guruge, *Return to Righteousness*, p. 442.

80 Ibid., p. 43.

81 Ibid., p. 443.

82 Ibid., pp. 456–57.

83 Ibid., p. 566.

84 Ibid., p. 155.

85 Ibid., p. 465.

86 Ibid., p. 439.

87 Ibid., p. 452.

88 Ibid., p. 440.

89 Ibid., p. 450.

90 Ibid., p. 499.

91 Ibid., p. 500.

92 Ibid., p. 465.

93 Ibid., p. 78.

94 Ibid., p. 243.

95 Ibid., p. 741.

96 Bhāvaviveka, *Prajñāpradīpamūlamadhyamikavṛtti*, in *Sde dge Tibetan Tripiṭaka Bstan Ḥgyur Preserved at the Faculty of Letters, University of Tokyo*, (Dbu ma) 2, 199 (Tsha) (Tokyo: Sekai Seiten Kanko Kyokai, 1977), 46a1–5.

97 Avalokitavrata, *Prajñāpradīpaṭīkā*, in ibid., (Dbu ma) 4, 201 (Wa), 15a3–15b3. This passage seems to have been first noticed by Yūichi Kajiyama. See his "Bhāvaviveka's Prajñāpradīpaḥ (1. Kapitel)," *Wiener Zeitschrift für die Kunde Süd- und Ostasiens* 7 (1963): 40–41. The first English-language reference (which refers to Kajiyama's essay and not Avalokitavrata's text) occurs in J. W. de Jong, "Buddhism and the Equality of the Four Castes," in *A Green Leaf: Papers in Honour of Professor Jes P. Asmussen*, ed. W. Sundermann, J. Duchesne-Guillemin, and F. Vahman, Acta Iranica 28 (Leiden: E. J. Brill, 1988), pp. 429–30. De Jong's one-page summary of this paper, with the same title, appeared in *Earliest Buddhism and Madhyamaka*, ed. David Seyfort Ruegg and Lambert Schmidthausen (Leiden: E. J. Brill, 1990), p. 58.

Chapter 3

1 In Tibetan literature of the eighteenth century, we find guidebooks to Shambhala and to the pilgrimage places in the Kathmandu valley, but little about India until 1789 and a work called *Discourse on India to the South, a Mirror of the Eight Topics of Analysis* (*Lho phyogs rgya gar gyi gtam brtag pa brgyad kyi me long*) by the prominent Nyingma lama 'Jigs med gling pa (1729–1798). (The eight topics of analysis are gems, lands, garments, trees, horses, elephants, men, and women.) 'Jigs med gling pa never

went to India himself, but one of his disciples did, the Bhutanese monk and government official Byang chub rgyal mtshan, who served as emissary to the East India Company in Calcutta for three years, circa 1775. He seemed to have traveled widely during his stay and reported his experiences to his teacher upon his return, on the basis of which 'Jigs med gling pa composed the *Discourse on India to the South.* Like so much travel literature of the seventeenth and eighteenth centuries, the work is a combination of fantasy, history, and marvels. For example, Byang chub rgyal mtshan describes "the [tribes] of the *Klo*[-pa] called Khaptra (*Kha-khra*) and Gidu (*Ghri-dho*) whose sons cut off the heads of their mothers as wedding gifts for their brides when they get married." See Michael Aris, *'Jigs-med-gling-pa's "Discourse on India" of 1789: A Critical Edition and Annotated Translation of the Lho-phyogs rgya-gar-gyi gtam brtag-pa brgyad-kyi me-long* (Tokyo: International Institute for Buddhist Studies of ICABS, 1996), p. 19.

Yet he understands British maritime trade quite well.

> At the point where [that river] arrives at Calcutta (*Ka-li-ka-ta*), the British (*Phe-reng-ba*, Ferengi) take to round ships which are not fixed with iron nails but instead bound with so-called *tshar* (strips) of bamboo twenty spans in length, and they depart for trade to China (*rGya-nag*). Although the Chinese (*rGya-nag-pa*) do not let them proceed to Beijing (*Pi-cing*), they have given them a place [at Canton] for the transaction of trade, and so at that place there is a meeting of both India and China. On account of that it is said that in India there are all the products of China (*Tsi-na*). (Ibid., p. 23)

And he was an acute observer, writing of the British, whom he calls Ferengis.

> Since they are expert in the play of crafts, [they are able to make] a box within which there are lots of large and small pipes made of tin [for producing] sounds that are both loud and soft, and many kinds of hollow tubes of various shapes needed for various resonances. . . . By winding a screw on the outside a wind is produced from a bellows inside made of soft leather, emitting a very sweet-sounding tune like a human voice issuing forth. (Ibid., p. 57)

2 For a translation and study of his *Adornment for Nāgārjuna's Thought* (*Klu sgrub dgongs rgyan*), see Donald S. Lopez Jr., *The Madman's Middle Way: Reflections on Reality of the Tibetan Monk Gendun Chopel* (Chicago: University of Chicago Press, 2005).

3 Hor khang bsod nams dpal 'bar, ed., *Dge 'dun chos 'phel gyi gsung rtsom*, vol. 2 (Gangs can rig mdzod 11) (Lhasa, Tibet: Bod ljongs bod yig dpe rnying dpe skrun khang, 1990), p. 166.

4 Ibid., pp. 166–173. Gendun Chopel's essay on science is discussed briefly in Thupten Jinpa, "Science as an Ally or a Rival Philosophy? Tibetan Buddhist Thinkers' Engagement with Modern Science," in *Buddhism and Science*, ed. B. Alan Wallace (New York: Columbia University Press, 2003), pp. 71–85, especially pp. 71–75.

5 Buddhadasa P. Kirthisinghe, ed., *Buddhism and Science*, reprint ed. (Delhi: Motilal Banarsidass, 1996), p. 4.

6 Hor khang bsod nams dpal 'bar, ed. *Dge 'dun chos 'phel gyi gsung rtsom*, 2:68.

7 Rdo rje rgyal, *'Dzam gling rig pa'i dpa' bo rdo brag dge 'dun chos 'phel gyi byung ba brjod pa bden gtam rna ba'i bcud len* (Gansu, China: Kan su'u mi rigs dpe skrung khang, 1997), pp. 66–67. Translation is from Gendun Chopel, *In the Forest of Faded Wisdom: 104 Poems by Gendun Chopel, a Bilingual Edition*, ed. and trans. Donald S. Lopez Jr. (Chicago: University of Chicago Press, forthcoming). Copyright © 2009 by The University of Chicago. All rights reserved.

8 Hor khang bsod nams dpal 'bar, *Dge 'dun chos 'phel gyi gsung rtsom*, 2:156.

9 Ibid., 2:161.

10 Ibid., 2:411. Translation is from Chopel, *In the Forest of Faded Wisdom*. See note 7 above.

11 His Holiness the Dalai Lama, *The Universe in a Single Atom: The Convergence of Science and Spirituality* (New York: Morgan Road Books, 2005). Some of the other recent works in which the Dalai Lama discusses Buddhism and Science include Anne Harrington and Arthur Zajonc, eds., *The Dalai Lama at MIT* (Cambridge, MA: Harvard University Press, 2006); Arthur Zajonc, ed., *The New Physics and Cosmology: Dialogues with the Dalai Lama* (Oxford: Oxford University Press, 2004); Jeremy W. Hayward and Francisco Varela, eds., *Gentle Bridges: Conversations with the Dalai Lama on the Sciences of Mind* (Boston: Shambhala, 2001); Richard J. Davidson and Anne Harrington, eds., *Visions of Compassion: Western Scientists and Tibetan Buddhists Examine Human Nature* (Oxford: Oxford University Press, 2002); and Zara Houshmand, Robert B. Livingston, and B. Alan Wallace, eds., *Consciousness at the Crossroads: Conversations with the Dalai Lama on Brain Science and Buddhism* (Ithaca, NY: Snow Lion Publications, 1999).

12 His Holiness the Dalai Lama, *The Universe in a Single Atom*, p. 2.

13 Ibid., pp. 22–23.

14 Ibid., p. 4.

15 The Dalai Lama, *The Way to Freedom* (San Francisco: HarperSanFrancisco, 1994), p. 73.

16 His Holiness the Dalai Lama, *The Universe in a Single Atom*, p. 3.

17 Ibid., p. 50.

18 Ibid., p. 84.

19 Ibid., p. 25.

20 *Saṃyutta Nikāya* 12.65. Translation (with slight adaptations) from *The Connected Discourses of the Buddha: A New Translation of the Saṃyutta Nikāya*, trans. Bhikkhu Bodhi (Boston: Wisdom Publications, 2000), 1:603–4.

21 Ibid., p. 24.

22 For a trenchant analysis of the role of the category of experience in Buddhist thought and practice, see Robert H. Sharf, "Buddhist Modernism and the Rhetoric of Meditative Experience," *Numen* 42 (1995): 228–83. Sharf argues that in classical Buddhist texts on meditation practice, including Kamalaśīla's *Bhāvanākrama*, Buddhaghosa's *Visuddhimagga*, Zhiyi's *Mohe zhiguan*, and Tsong kha pa's *Lam rim chen mo*,

the author describes meditative practice and its resultant effects through appeal to scripture, and not to the author's own experience. The rhetoric of experience, he argues, plays a much more significant role in various forms of Buddhist modernism, including the *vipassanā* movement and certain forms of Zen imported to the West.

In *The Universe in a Single Atom*, the Dalai Lama writes, "The Buddhist understanding of mind is primarily derived from empirical observations grounded in the phenomenology of experience, which includes the contemplative techniques of meditation" (p. 135). The Dalai Lama thus implies that meditative experience is the source of Buddhist epistemology. Some thirty years ago, there was a spirited discussion of the relationship between religious doctrine and religious experience. In the case of Buddhism, Robert Gimello argued that, "rather than speak of Buddhist doctrines as interpretations of Buddhist mystical experiences, one might better speak of Buddhist mystical experiences as deliberately contrived exemplifications of Buddhist doctrine." See Robert Gimello, "Mysticism and Meditation," in *Mysticism and Philosophical Analysis*, ed. Steven Katz (New York: Oxford University Press, 1978), p. 193.

23 In some of the tantric traditions, however, this is not the case. The Buddha must leave his physical body behind under the Bodhi tree and in a mind-made body go to the highest heaven, where he receives tantric initiation from the buddhas of the ten directions. He then returns to Bodh Gayā and achieves enlightenment. See David Snellgrove, *Indo-Tibetan Buddhism: Indian Buddhists and Their Tibetan Successors* (Boston: Shambhala, 1987), 1:119–21.

24 Henry Clarke Warren, *Buddhism in Translations: Passages Selected from the Buddhist Sacred Books and Translated from the Original Pali into English* (Cambridge, MA: Harvard University Press, 1953), p. 14. For a more modern translation, see N. A. Jayawickrama, trans., *The Story of Gotama Buddha (Jātaka-nidāna)* (Oxford: Pali Text Society, 2002), p. 18.

25 For examples of some of the interpretive strategies employed in various Buddhist traditions, see Donald S. Lopez Jr., ed., *Buddhist Hermeneutics* (Honolulu: University of Hawai'i Press, 1988).

26 For a translation and commentary of a standard Dge lugs textbook on this topic, see Lati Rinbochay, *Mind in Tibetan Buddhism*, trans., ed., and with an introduction by Elizabeth Napper (Ithaca, NY: Snow Lion Publications, 1980), especially pp. 75–82. For a useful discussion of the inference based on scripture and some of the problems that attend the category, see Tom J. F. Tillemans, "How Much of a Proof Is Scripturally Based Inference?" in his *Scripture, Logic, Language: Essays on Dharmakīrti and His Tibetan Successors* (Boston: Wisdom Publications, 1999), pp. 37–51. See also pp. 27–30.

27 His Holiness the Dalai Lama, *The Universe in a Single Atom*, p. 28.

28 In a similar vein, Dharmakīrti states at *Pramāṇavārttika* 1.217, "Because it is certain that he possesses the method for that which is to be adopted and discarded, he is not deceptive regarding the primary aim. Thus, one can infer [that he is reliable] regarding other things." The phrase "that which is to be adopted and discarded" is often interpreted to refer to the four truths, where the truths of suffering and origin are to be discarded and the truths of cessation and path are to be adopted. Thus, the passage means that because the Buddha is correct about important matters such as the four truths (which fall into the category of the hidden and thus can be confirmed by reasoning), it can be inferred that he is correct about other things, including those that fall into the category of the very hidden. For a useful study of a Tibetan text on the authority of the Buddha, see Tom J. F. Tillemans, *Persons of Authority* (Stuttgart, Germany: Franz Steiner Verlag, 1993).

29 *Tattvasaṃgraha*, śloka 3588. For the Sanskrit, see Embar Krishnamacharya, ed., *Tattvasaṅgraha of Śāntarakṣita with the Commentary of Kamalaśīla* (Varanasi, India: Bauddha Bharati, 1968), 2:922. This edition also contains Kamalaśīla's *Tattvasaṃgrahapañjikā*. For an English translation of both works, see Ganganatha Jha, trans., *The Tattvasaṅgraha of Shāntarakṣita with the Commentary of Kamalashīla*, 2 vols., reprint of 1938 edition (Delhi: Motilal Banarsidass, 1986).

Since the nineteenth century, the Pāli text most commonly cited in support of the empirical approach of Buddhism is the *Kālāma Sutta* of the *Aṅguttara Nikāya* (3.65). The relevant passage occurs when the Kālāmas ask the Buddha:

> "There are, Lord, some ascetics and Brahmins who come to Kesaputta. They explain and elucidate their own doctrines, but disparage, debunk, revile and vilify the doctrines of others. But then some other ascetics and Brahmins come to Kesaputta, and they too explain and elucidate their own doctrines, but disparage, debunk, revile and vilify the doctrines of others. For us, Lord, there is perplexity and doubt as to which of these good ascetics speak truth and which speak falsehood."
>
> [The Buddha replies:] "It is fitting for you to be perplexed, O Kālāmas, it is fitting for you to be in doubt. Doubt has arisen in you about a perplexing matter. Come, Kālāmas. Do not go by oral tradition, by lineage of teaching, by hearsay, by a collection of scriptures, by logical reasoning, by inferential reasoning, by reflection of reasons, by the acceptance of a view after pondering it, by the seeming competence of a speaker, or because you think, 'The ascetic is our teacher.' But when you know for yourselves, 'These things are unwholesome, these things are blamable; these things are censured by the wise; these things, if taken and practiced, lead to harm and suffering,' then you should abandon them. . . . But when you know for yourselves, 'These things are wholesome, these things are blameless; these things are praised by the wise; these things, if undertaken and practiced, lead to welfare and happiness,' then you should engage in them."

The Buddha then goes on to set forth the virtues of the meditative practice of love, compassion, joy, and equanimity. See Nyanaponika Thera and Bhikkhu Bodhi, *Numerical Discourses of the Buddha: An Anthology of Suttas from the Aṅguttara Nikāya* (Walnut Creek, CA: AltaMira Press, 1999), pp. 65–66.

Alluding to this passage in an address entitled "Message of the Buddha" delivered at The Town Hall in New York in 1925, Dharmapāla declared, "The scientific basis of the Religion that the Buddha taught is made manifest by His approval of the scepticism which was visible in the Kalama community who declined to believe the dogmas of religion." See Ananda Guruge, ed., *Return to Righteousness: A Collection of Speeches, Essays and Letters of Anagarika Dharmapāla* (Ceylon: Government Press, 1965), p. 30. For thoughtful reconsiderations of this passage in the *Kālāma Sutta*, see Bhikkhu Bodhi, "A Look at the 'Kalama Sutta,'" *BPS* [Buddhist Publication Society] *Newsletter* 9 (Spring 1988): cover essay, and Martin Verhoeven, "Western Science, Eastern Spirit: Historical Reflections on the East/West Encounter," *Religion East & West: Journal of the Institute for World Religions* 3 (June 2003): 46–48.

Typical of more recent declarations of Buddhism's apparent absence of dogmatism is, "Buddhism stands ready to revise its beliefs at any moment if they are proved to be wrong." But the author immediately qualifies this bold claim: "Not that it has any doubts about the basic truth of its discoveries, nor does it expect that the results it has built up over 2,500 years of contemplative science will suddenly be invalidated." See Matthieu Ricard and Trinh Xuan Thuan, *The Quantum and the Lotus: A Journey to the Frontiers Where Science and Buddhism Meet* (New York: Three Rivers Press, 2001), p. 3.

30 *Tattvasaṃgraha* 3268–76.

31 *Tattvasaṃgraha* 3322–44.

32 *Tattvasaṃgraha* 3368–69.

33 *Tattvasaṃgraha* 3550–51.

34 *Tattvasaṃgraha* 3586–88. For the Sanskrit, see Krishnamacharya, *Tattvasaṅgraha of Śāntarakṣita*, 2:922. For Jha's translation, see *The Tattvasaṅgraha of Shāntarakṣita*, 2:1558.

35 *Tattvasaṃgraha* 3606–11.

36 *Tattvasaṃgraha* 3627–29.

37 His Holiness the Dalai Lama, *The Universe in a Single Atom*, p. 25.

38 Thomas W. Rhys Davids, *Lectures on the Origin and Growth of Religion as Illustrated by Some Points in the History of Indian Buddhism*, Hibbert Lectures 1881 (London: Williams and Norgate, 1881), p. 94. Claims of the compatibility of Buddhism and evolution were common in the Victorian and Edwardian periods. The *Frankfurter Zeitung* reported on April 25, 1890, "In itself Buddhism is a profound and all-embracing doctrine, adapted particularly for our time, because it does not contradict science, but contains on the contrary, the germs of scientific truths. For example, Transformism or Darwinism is involved in Buddhism." See *The Literary Digest*, May 31, 1890, p. 22 (162). We read in a 1905 essay in *The Fortnightly Review*, "The religion of the Buddha is not in conflict with modern science; he anticipated many of its most important conclusions; its primary principle of evolution is one with his central tenet." See W. S. Lilly, "The Message of Buddhism to the Western World," *The Fortnightly Review*, n.s., 78 (July–December 1905): 213.

39 Thomas H. Huxley, *Evolution and Ethics and Other Essays* (London: Macmillan and Company, 1894), p. 61.

40 His Holiness the Dalai Lama, *The Universe in a Single Atom*, p. 112.

41 Ibid., p. 131.

42 Ibid., p. 131.

43 Ibid., pp. 110, 111. Earlier in the book (on page 12), the Dalai Lama makes a similar comment, where he is apparently drawing a parallel between what he calls "scientific materialism" and Intelligent Design: "The view that all aspects of reality can be reduced to matter and its various particles is, to my mind, as much a metaphysical position as the view that an organizing intelligence created and controls reality."

44 Ibid., p. 13.

Chapter 4

1 H. Fielding, *The Soul of a People* (London: Richard Bentley and Son, 1898), p. 22.

2 Ibid., p. 167.

3 Ibid., p. 23.

4 The following is the full list of the "Fundamental Buḍḍhistic Beliefs" as they appear in the appendix to the forty-fourth edition of Olcott's *The Buḍḍhist Catechism* (with Olcott's idiosyncratic diacritical marks retained):

> I. Buḍḍhists are taught to show the same tolerance, forbearance, and brotherly love to all men, without distinction; and an unswerving kindness towards the members of the animal kingdom.
> II. The universe was evolved, not created; and it functions according to law, not according to the caprice of any God.
> III. The truths upon which Buḍḍhism is founded are natural. They have, we believe, been taught in successive kalpas, or world periods, by certain illuminated beings called BUḌḌHAS, the name BUḌḌHA meaning "Enlightened."
> IV. The fourth teacher in the present kalpa was Sākya Muni, or Gauṭama Buḍḍha, who was born in a royal family in India about 2,500 years ago. He is an historical personage and his name was Siḍḍhārṭha Gauṭama.
> V. Sākya Muni taught that ignorance produces desire, unsatisfied desire is the cause of rebirth, and rebirth the cause of sorrow. To get rid of sorrow, therefore, it is necessary to escape rebirth; to escape rebirth, it is necessary to extinguish desire; and to extinguish desire, it is necessary to destroy ignorance.
> VI. Ignorance fosters the belief that rebirth is a necessary thing. When ignorance is destroyed the worthlessness of every such rebirth, considered as an end in itself, is perceived, as well as the paramount need of adopting a course of life by which the necessity for such repeated births can be abolished. Ignorance also begets the illusive and illogical idea that there is only one existence for man, and the other illusion that this one life is followed by states of unchangeable pleasure or torment.
> VII. The dispersion of all this ignorance can be attained by the persevering practice of an all-embracing altruism in conduct, development of intelligence, wisdom in thought, and destruction of desire for the lower personal pleasures.
> VIII. The desire to live being the cause of rebirth, when that is extinguished rebirths cease and the perfected individual attains by meditation that highest state of peace called *Nirvāṇa*.
> IX. Sakya Muni taught that ignorance can be dispelled and sorrow removed by the knowledge of the four Noble Truths, *viz*:

1. The miseries of existence;
2. The cause productive of misery which is the desire ever renewed of satisfying oneself without being able ever to secure that end;
3. The destruction of that desire, or the estranging of oneself from it;
4. The means of obtaining this destruction of desire. The means which he pointed out is called the Noble Eightfold Path, *viz*: Right Belief; Right Thought; Right Speech; Right Action; Right Means of Livelihood; Right Exertion; Right Remembrance; Right Meditation.

x. Right Meditation leads to spiritual enlightenment, or the development of that Buḍḍha-like faculty which is latent in every man.
xi. The essence of Buḍḍhism summed up by the Taṭhāgaṭhā (Buḍḍha) himself is:

To cease from all sin,
To get virtue,
To purify the heart.

xii. The universe is subject to a natural causation known as "Karma". The merits and demerits of a being in past existences determine his condition in the present one. Each man, therefore, has prepared the causes of the effects which he now experiences.
xiii. The obstacles to the attainment of good karma may be removed by the observance of the following precepts, which are embraced in the moral code of Buḍḍhism, *viz*: (1) Kill not; (2) Steal not; (3) Indulge in no forbidden sexual pleasure; (4) Lie not; (5) Take no intoxication or stupefying drug or liquor. Five other precepts, which need not here be enumerated should be observed by those who would attain, more quickly than the average layman, the release from misery and rebirth.
xiv. Buḍḍhism discourages superstitious credulity. Gauṭama Buḍḍha taught it to be the duty of a parent to have his child educated in science and literature. He also taught that no one should believe what is spoken by any sage, written in any book, or affirmed by a tradition, unless it accord with reason.

See Henry S. Olcott, *The Buddhist Catechism*, 44th ed. (Adyar, India: Theosophical Publishing House, 1915), pp. 92–95.

This was not the only attempt to capture the fundamentals of Buddhism in the form of a list. In 1890 the Jōdo Shinshū Buddhist priest and scholar Murakami Senshō (1851–1929) presented a list of the "Ten Features Pervading Buddhist Teachings" in a work entitled *Bukkyō ikkan ron* (*The Consistency of Buddhism*).

1. The totality of existing entities (*ban'yū*) is boundless and limitless, both horizontally (the spatial dimension) and vertically (the temporal dimension).

2. The totality of phenomena (*banshō*) includes three major laws: causality, impermanence, and egolessness.
3. The essence (*hontai*) of the totality of existing entities is unborn, undying, it does not increase nor decrease, it is equal (*byōdō*) and without differentiation.
4. There are two major approaches in Buddhist philosophy: the model of dependent origination (*engiron*) and that of the real state (*jissōron*).
5. The model of dependent origination in the Buddhist teachings is spatial and not temporal.
6. The model of dependent origination in the Buddhist teachings is subjective and not objective.
7. Buddhist teachings consider the three aspects of ethical conduct, contemplation, and wisdom as the basis for engaging in practice.
8. Buddhist teachings require that one abandon all deluding attachments and break away from all defilements.
9. Buddhist teachings consider leaving the deluded state and reaching the awakened state as the goal.
10. There are two major approaches for reaching the awakened state, arduous practice (*nangyō*) and easy practice (*igyō*).

See Michel Mohr, "Murakami Senshō: In Search for the Fundamental Unity of Buddhism," *The Eastern Buddhist* 37, nos. 1 and 2 (2006): 10–11.

5 Henry Steel Olcott, *Old Diary Leaves: The Only Authentic History of the Theosophical Society*, 3rd ser. (1887–92), 2nd ed. (Adyar, India: Theosophical Publishing House, 1931), pp. 62–64. Olcott provides a briefer version of exchange in the previous volume of *Old Diary Leaves*:

> "You have done nobly," he [Müller] said, "in helping so much to revive the love for Sanskrit, and the Orientalists have watched the development of your Society with the greatest interest from the commencement. But why will you spoil all this good reputation by pandering to the superstitious fancies of the Hindus, by telling them that there is an esoteric meaning in their Shastras? I know the language perfectly, and I assure you there is no such thing as a Secret Doctrine in it." In reply, I simply told the Professor that every unspoilt (i.e., unwesternized) Pandit throughout all India believed, as we did, in the existence of this hidden meaning; and that, as for the *Siddhis*, I personally knew men who possessed them and whom I had seen exhibit their powers. "Well, then," said my erudite host, "let us change the subject."

See Henry Steel Olcott, *Old Diary Leaves: The Only Authentic History of the Theosophical Society*, 3rd ser. (1883–87), 2nd ed. (Adyar, India: Theosophical Publishing House, 1929), p. 177.

6 *The Life and Letters of the Right Honourable Friedrich Max Müller Edited by His Wife*, 2 vols. (London: Longmans, Green, and Co., 1902), 2:234.

7 See *The Book of Ser Marco Polo the Venetian Concerning the Kingdoms and Marvels of the East*, 2 vols., trans. and ed. Sir Henry Yule, 3rd ed., revised by Henri Cordier

(New York: AMS, 1986), reprint of 1926 London edition, 2:316–17. See also the extensive notes of Yule and Cordier, pp. 320–30.

8 Ibid., p. 318.

9 Denis Diderot and Jean d'Alembert, *Encyclopédie, ou Dictionnaire raisonné des sciences, des métiers et des arts* (Paris, 1765), s.v. "Siaka."

10 Max Müller, *Chips from a German Workshop*, vol. 1, *Essays on the Science of Religion*, reprint (Chico, CA: Scholars Press, 1985), pp. 196–97.

11 "Sur quelques épithètes descriptives de Bouddha qui font voir que Bouddha n'appartenait pas à la race nègre," in Jean-Pierre Abel-Rémusat, *Mélanges Asiatiques* (Paris: Librarie Orientale de Dondey-Dupré Père et Fils, 1825), 1:100–128.

12 William Erskine, "Account of the Cave-Temple of Elephanta with a Plan of the Drawings of the Principal Figures," *Transactions of the Literary Society of Bombay* (1819), 1:201, 202.

13 See Philip Almond, *The British Discovery of Buddhism* (Cambridge: Cambridge University Press, 1988), pp. 29–30.

14 Brian Houghton Hodgson, "Quotations from Original Sanskrit Authorities in proof and illustration of the preceding article," *Journal of the Asiatic Society of Bengal* 5 (1836): 67–68. Writing in 1844, Eugène Burnouf offered a similar view:

> The distinctive character of Buddhism appears here, a doctrine where moral practice dominates, and which is distinguished in this way from Brahmanism, where philosophical speculation on the one hand, and mythology on the other, certainly occupy a greater place. In that as well, Buddhism testifies clearly to its posteriority to Brahmanism. If moral systems indeed are only born following ontological systems, which is established in the most definitive manner by the history of Greek philosophy, then Buddhism must necessarily, and if one can express oneself in this way, genetically be posterior to Brahmanism. Certainly the elements of Brahmanical science are not exclusively ontological, and the study of moral man already appears there, but speculative research is nonetheless the dominant principle that gives a uniform direction to the whole of Brahmanism.

Eugène Burnouf, *Introduction à l'histoire du Buddhisme indien* (Paris: Imprimerie Royale, 1844), pp. 335–36. This and all subsequent translations from Burnouf are from Eugène Burnouf, *Introduction to the History of Indian Buddhism*, trans. Katia Buffetrille and Donald S. Lopez Jr. (Chicago: University of Chicago Press, forthcoming).

15 Brian H. Hodgson, "Sketch of Buddhism, derived from Bauddha Scriptures of Nipál," *Transactions of the Royal Asiatic Society of Great Britain and Ireland* (1830), 2:222–23.

16 Brian H. Hodgson, "Notice of the Languages, Literature, and Religion of Nepál and Tibet," *Asiatic Researches* 16 (1828).

17 For a more detailed examination of Hodgson's contributions to Buddhist studies, see, as the title suggests, Donald S. Lopez Jr., "The Ambivalent Exegete: Hodgson's Contributions to Buddhist Studies," in *The Origins of Himalayan Studies:*

Brian Houghton Hodgson in Kathmandu and Darjeeling, 1820 to 1858, ed. David Waterhouse (London: Routledge Curzon, 2004), pp. 49–76.

18 Upon receiving the Tibetan canon that Hodgson dispatched to Calcutta, Henry Thoby Prinsep wrote to him in a letter dated August 6, 1835, "This is indeed glorious and will redound to your immortal fame. I have told Csoma [de Körös] that he must on no account run away until he has read the whole of the Stangyur and made known its contents." The letter is preserved in Hodgson's "autograph book" at the Royal Asiatic Society in London.

19 Eduard Roer, review of *Introduction à l'histoire du Buddhisme indien*, by Eugène Burnouf, *Journal of the Asiatic Society of Bengal* 14, no. 2 (1845): 783.

20 Burnouf, *Introduction*, p. 11.

21 Ibid., p. 31.

22 In 1844 Thoreau published "The Preaching of the Buddha" in Emerson's Transcendentalist journal *The Dial*. It included a translation from the French by Elizabeth Palmer Peabody of a chapter from the *Lotus Sūtra*, drawn from two articles published by Burnouf in the previous year. The translation itself is often mistakenly attributed to Thoreau. See Roger C. Mueller, "A Significant Buddhist Translation by Thoreau," *The Thoreau Society Bulletin* (Winter 1977): 1–2.

23 Schopenhauer purchased the book the year after it was published. He cited it in the second edition of *The World as Will and Representation* and recommended it in his *On the Will in Nature*.

24 Quoted in Raymond Schwab, *The Oriental Renaissance: Europe's Rediscovery of India and East, 1680–1880* (New York: Columbia University Press, 1984), p. 439.

25 Burnouf, *Introduction*, p. 527.

26 Ibid., pp. 152–53.

27 Ibid., p. 338.

28 Ibid., p. 348.

29 Ibid., p. 194.

30 Ibid., pp. 126–27.

31 See "European Speculations on Buddhism," reprinted in Brian Houghton Hodgson, *Essays on the Languages, Literature, and Religion of Nepal and Tibet: Together with Further Papers on the Geography, Ethnology, and Commerce of Those Countries* (London: Trubner and Company, 1874; reprint, New Delhi: Asian Educational Services, 1991), p. 100. For an examination of the tensions between the vernacular and the textual in another Buddhist setting, Sri Lanka, in a somewhat later period, see Charles Hallisey, "Roads Taken and Not Taken in the Study of Theravāda Buddhism," in *Curators of the Buddha: The Study of Buddhism under Colonialism*, ed. Donald S. Lopez Jr. (Chicago: University of Chicago Press, 1995), pp. 31–61.

32 *The Life and Letters of the Right Honourable Friedrich Max Müller*, 2:290.

33 F. Max Müller, "Esoteric Buddhism," *The Nineteenth Century: A Monthly Review* 33 (January–June 1893): 769. The essays (Müller's essay, Sinnett's reply, and Müller's rejoinder) were reprinted in F. Max Müller, *Last Essays, Second Series: Essays on the Science of Religion* (New York: Longmans, Green, and Co., 1901), pp. 79–170. For a scathing essay on the "religionette" of Esoteric Buddhism, published eight years before Müller's, see Frederika MacDonald, "Buddhism and Mock Buddhism," *Fortnightly Review*, n.s., 37 (January 1–June 1, 1885): 703–16.

34 Müller, "Esoteric Buddhism," p. 769.

35 Ibid., p. 773.

36 Ibid., p. 774.

37 Ibid., p. 775.

38 Ibid., p. 781.

39 Ibid., p. 784.

40 Ibid., p. 786.

41 Ibid., p. 787.

42 A. P. Sinnett, "Esoteric Buddhism (A Reply to Professor Max Müller)," *The Nineteenth Century: A Monthly Review* 33 (January–June 1893): 1015.

43 Ibid., pp. 1015–16.

44 Ibid., p. 1018.

45 Ibid.

46 Ibid., p. 1020.

47 Ibid., p. 1021.

48 Ibid., p. 1026.

49 Ibid., pp. 1026–27.

50 F. Max Müller, "Esoteric Buddhism: A Rejoinder," *The Nineteenth Century: A Monthly Review* 34 (July–December 1893): 297. The statement of Sinnett's that Müller paraphrases is: "But whether I obtained the teaching on which *Esoteric Buddhism* rests from a Mahâtma on the other side of the Himalayas or evolved them out of my own head need only interest people who begin to be seriously interested in the teaching on its own *primâ facie*, intrinsic claims." See Sinnett, "Esoteric Buddhism (A Reply to Professor Max Müller)," p. 1022.

51 Müller, "Esoteric Buddhism: A Rejoinder," p. 303.

52 *The Mahatma Letters to A. P. Sinnett from the Mahatmas M. & K. H.*, transcribed, compiled, and with an introduction by A. T. Barker (London: Rider and Company, 1948), p. 185.

53 Ibid., p. 158.

54 Ibid., p. 110.

55 K. Paul Johnson, *The Masters Revealed: Madame Blavatsky and Myth of the Great White Lodge* (Albany: State University of New York Press, 1994), pp. 148–75.

56 Henry Steel Olcott, *Old Diary Leaves: America 1874–1878* [also referred to as part 1] (Adyar, India: Theosophical Publishing House, 1941), p. 396.

57 Ibid., p. 406.

58 Allen Ginsberg, "Sakyamuni Coming Out from the Mountain," in *Reality Sandwiches* (San Francisco: City Lights Books; first published 1963), pp. 9–10.

59 Burnouf, *Introduction*, pp. 336–37.

60 *The Life and Letters of the Right Honourable Friedrich Max Müller*, 2:297.

61 Müller, "Esoteric Buddhism," p. 783.

62 Müller, "Esoteric Buddhism: A Rejoinder," p. 303.

63 A. P. Sinnett, *Esoteric Buddhism* (Boston: Houghton, Mifflin and Co., 1895), pp. 209–10.

64 Ibid., p. 220.
65 Ibid.
66 Ibid., p. 221.
67 For a brief description of Blavatsky's race theory, see Donald S. Lopez Jr., *Prisoners of Shangri-La: Tibetan Buddhism and the West* (Chicago: University of Chicago Press, 1998), pp. 54–55.
68 See ibid., pp. 46–71.
69 See Dharmapāla, "Can a Buddhist Be a Member of the Theosophical Society?" *Maha-Bodhi and the United Buddhist World*, 14, no. 3 (March 1906): 42. In 1907 he wrote:

> The Dharma stands unique on its own basis and it has no connection with the latter "isms" founded by enthusiasts and visionaries. Theosophy as a pantheism may claim kinship with other existing religions which posit a creator and a selfish ego. The Dharma of the Tathāgata, we emphasize, has no relationship with any other existing religion. It repudiates a creator, has no sympathy with revelations, and acknowledges no mahatmic authority. Dharma alone is supreme to the Buddhist.

See Dharmapāla, "Colonel Olcott and the Buddhist Revival Movement," *Maha-Bodhi and the United Buddhist World* 15, nos. 1–3 (January–March 1907): 28.

In the same issue of *Maha-Bodhi*, Dharmapāla included the following in his obituary of Olcott, his old friend and patron:

> Although he became a Buddhist he does not seem to have grasped the fundamentals of Buddhism. In his Catechism made for the use of beginners in Buddhism he introduced many Brahmanical ideas; such as the existence of a universal over soul. He formed a false conception of Nirvana, classing the desire for the attainment of Nirvana in the same category as the desire for heavenly bliss.
>
> He hurt the susceptibilities of the Buddhists by his remarks on the humanity of the Buddha and the sacred Tooth-relic still existant [*sic*] in the Dalada Maligawa at Kandy. His exact position in the Buddhist revival movement is also difficult to judge, the tendency being to give him too much credit rather than too little. . . .
>
> If Colonel Olcott looked upon Buddhism as the daughter-religion of Hinduism he has done a great dishonour to Buddhism.
>
> We judge only his views; not himself personally. Of him we have nothing but the kindest love; of the dead nothing but good.

See Dharmapāla, "Obituary," *Maha-Bodhi and the United Buddhist World* 15, nos. 1–3 (January–March 1907): 19, 20. In the same issue, we find a piece entitled "What Buddha Thought of Occultism." In an earlier piece entitled "The Parting of the Ways," Dharmapāla writes, "This trend of circumstances which had been foreshadowed and feared by some of the earlier learned Bhikkhus has at present become a menace, and the pure Doctrines of the Blessed One are in danger of being mixed with the Theosophical leaven." See *Maha-Bodhi and the United Buddhist World* 14, no. 8 (August 1906): 118. Elsewhere, in an essay entitled "Theosophical Degenerates," he writes, "Theosophy is a Eurasian pantheism, it is neither purely Eastern nor Western. It is an 'occult' mixture given to the credulous world by a band of impostors, who deceive the world by plagiarising Buddhistic Pali phrases and Vedantic metaphysics." See *Maha-Bodhi and the United Buddhist World* 14, no. 7 (July 1906): 106.

70 Ananda Guruge, ed., *Return to Righteousness: A Collection of Speeches, Essays and Letters of Anagarika Dharmapala* (Colombo: Ministry of Education and Cultural Affairs, 1965), p. 775.

71 Ibid., p. 20.

72 Ibid., p. 465.

73 Ibid., p. 439.

74 Ibid., p. 741.

75 See G. P. Malalasekhara, *Dictionary of Pāli Proper Names* (Delhi: Munshiram Manoharlal, 1998), 1:294–305.

Chapter 5

1 This description of a meditation session on Vajrayoginī is drawn from the following sources. Sermey Khensur Lobsang Tharchin, *Sublime Path to the Kechara Paradise: Vajrayogini's Eleven Yogas of Generation Stage Practice* (Howell, NJ: Mahayana Sutra and Tantra Press, 1997); Geshe Ngawang Dhargyey, *Vajrayogini Sadhana and Commentary* (Dharamsala, India: Library of Tibetan Works and Archives, 2003); Geshe Kelsang Gyatso, *Guide to Dakini Land: The Highest Yoga Tantra Practice of Buddha Vajrayogini* (London: Tharpa Publications, 1999); Elizabeth English, *Vajrayoginī: Her Visualizations, Rituals and Forms* (Boston: Wisdom Publications, 2002); and John C. Huntington and Dina Bangel, *The Circle of Bliss: Buddhist Meditational Art* (Chicago: Serindia Publications, 2003). The list of infractions of the bodhisattva and tantric vows is drawn from Donald S. Lopez Jr., "A Rite for Restoring the Bodhisattva and Tantric Vows," in *Buddhism in Practice*, ed. Donald S. Lopez Jr. (Princeton, NJ: Princeton University Press, 1995).

2 The account of the meditation session given in the preceding paragraphs is fictional. For the results of a similar experiment, see Herbert Benson, John W. Lehmann, M. S. Malhotra, Ralph F. Goldman, Jeffrey Hopkins, and Mark D. Epstein, "Body Temperature Changes during the Practice of gTum-mo Yoga," *Nature* 295 (January 21, 1982): 234–36. For an account by a monk who participated in a similar experiment, see Venerable Lobsang Tenzin, "Biography of a Contemporary Yogi," *Chö Yang: The Voice of Tibetan Religion and Culture* 3 (1990): 110–11. For an account of problems encountered in an attempt to perform similar tests, see Zara Houshmand, Anne Harrington, Clifford Saron, and Richard J. Davidson, "Training the Mind: First Steps in a Cross-Cultural Collaboration in Neuroscientific Research," in *Visions of Compassion: Western Scientists and Tibetan Buddhists Examine Human Nature*, ed. Richard J. Davidson and Anne Harrington (Oxford: Oxford University Press,

2002), pp. 3–17. For a brief but important consideration of some of the ethical issues raised by such experiments, see José Ignacio Cabezón, "Buddhism and Science: On the Nature of the Dialogue," in *Buddhism and Science: Breaking New Ground*, ed. B. Alan Wallace (New York: Columbia University Press, 2003), pp. 36–39.

3 For an insightful account of the recent role of Asian religions, including Buddhism, in neurological research, see Anne Harrington, *The Cure Within: A History of Mind-Body Medicine* (New York: W. W. Norton and Company, 2008), pp. 205–42.

4 The text, which is found in the *Dīgha Nikāya*, is also called the *Sigālovāda Sutta*. For a translation, see Maurice Walshe, trans., *The Long Discourses of the Buddha: A Translation of the Dīgha Nikāya* (Boston: Wisdom Publications, 1995), pp. 461–69.

5 See Donald S. Lopez Jr., *The Madman's Middle Way: Reflections on Reality of the Tibetan Monk Gendun Chopel* (Chicago: University of Chicago Press, 2006), p. 69.

6 See *The Middle Length Discourses of the Buddha: A New Translation of the Majjhima Nikāya;* original translation by Bhikkhu Ñāṇamoli; translation edited and revised by Bhikku Bodhi (Boston: Wisdom Publications, 1995), p. 155.

7 Thus, we read in a survey of research on Buddhist meditation:

> The reason, in brief, is that unlike many contemplative traditions, Buddhist traditions tend to offer extensive, precisely descriptive and highly detailed theories about their practices in a manner that lends itself readily to appropriation into a neuroscientific context. This emphasis on descriptive precision stems from the central role that various forms of meditation play in Buddhist practice. That is, from the standpoint of nearly every Buddhist tradition, some type of meditative technique must be employed if one is to advance significantly on the Buddhist spiritual path, and since Buddhism initially developed in a cultural context where a wide range of such techniques were available, Buddhist theoreticians recognized the need to specify exactly the preferred techniques. Their analyses eventually develop into a highly detailed scholastic tradition known in Sanskrit as the Abhidharma—a type of Buddhist "psychology" that also includes discussions of epistemology, philosophy of language, the composition of the material world, and cosmology.

See Antoine Lutz, John D. Dunne, and Richard J. Davidson, "Meditation and the Neuroscience of Consciousness: An Introduction," in *The Cambridge Handbook of Consciousness*, ed. Philip David Zelazo, Morris Moscovitch, and Evan Thompson (Cambridge: Cambridge University Press, 2007), p. 503. The same essay compares the results of studies of various forms of meditation, including TM. See pp. 532–43.

8 From *The Embodied Mind*, one of the most insightful and influential works in the Buddhism and cognitive science genre: "We believe that the Buddhist doctrines of no-self and of nondualism that grew out of this method [that is, mindfulness meditation] have a significant contribution to make to the dialogue with cognitive science." Later in the same volume, the authors state, "It is said that emptiness is a natural discovery that one would make by oneself with sufficient mindfulness/awareness—natural but shocking." See Francisco J. Varela, Evan Thompson, and Eleanor Rosch, *The Embodied Mind: Cognitive Science and Human Experience* (Cambridge, MA: MIT Press, 1997), pp. 21 and 225.

Conclusion

1 On the disappearance of the dharma, see Jan Nattier, *Once Upon a Future Time: Studies in a Buddhist Prophecy of Decline* (Berkeley, CA: Asian Humanities Press, 1991).

2 One of the therapies that has been developed in recent years is MBSR—Mindfulness-Based Stress Reduction. For another example of this denuding of the Buddhist tradition, see Sam Harris, "Killing the Buddha," *Shambhala Sun* (March 2006): 73–75, where the author explains, "In many respects, Buddhism is very much like science. One starts with the hypothesis that using attention in the prescribed way (meditation), and engaging in or avoiding certain behaviors (ethics), will bear the promised result (wisdom and psychological well-being). This spirit of empiricism animates Buddhism to a unique degree. For this reason, the methodology of Buddhism, if shorn of its religious encumbrances, could be one of our greatest resources as we struggle to develop our scientific understanding of human subjectivity" (p. 74).

3 See Frederika MacDonald, "Buddhism and Mock Buddhism," *Fortnightly Review*, n.s., 37 (January 1–June 1, 1885): 716.

4 Sanskrit versions in both Hīnayāna and Mahāyāna sources add a third criterion: that the words not contradict "the way things are" (*dharmatām na vilomayati*). This requirement seems redundant, since a doctrine found to be in accordance with the sūtra and the vinaya, yet contradicting the *dharmatā*, would seem inappropriate.

5 On this topic, see, for example, Donald S. Lopez Jr., "On the Interpretation of the Mahāyāna Sūtras," in *Buddhist Hermeneutics*, ed. Donald S. Lopez Jr. (Honolulu: University of Hawai'i Press, 1988), pp. 47–70.

6 Stanislaw Schayer, "Precanonical Buddhism," *Archiv Orientalni* 7 (1935): 124.

7 *The Middle Length Discourses of the Buddha: A New Translation of the "Majjhima Nikāya,"* original translation by Bhikkhu Ñānamoli; translation edited and revised by Bhikkhu Bodhi (Boston: Wisdom Publications, 1995), p. 167.

INDEX